AUTOMATIC SPEECH RECOGNITION

THE KLUWER INTERNATIONAL SERIES IN ENGINEERING AND COMPUTER SCIENCE

VLSI, COMPUTER ARCHITECTURE AND DIGITAL SIGNAL PROCESSING

Consulting Editor

Jonathan Allen

Other books in the series:

Logic Minimization Algorithms for VLSI Synthesis. R.K. Brayton, G.D. Hachtel, C.T. McMullen, and A.L. Sangiovanni-Vincentelli. ISBN 0-89838-164-9.
Adaptive Filters: Structures, Algorithms, and Applications. M.L. Honig and D.G. Messerschmitt. ISBN 0-89838-163-0.
Introduction to VLSI Silicon Devices: Physics, Technology and Characterization. B. El-Kareh and R.J. Bombard. ISBN 0-89838-210-6.
Latchup in CMOS Technology: The Problem and Its Cure. R.R. Troutman. ISBN 0-89838-215-7.
Digital CMOS Circuit Design. M. Annaratone. ISBN 0-89838-224-6.
The Bounding Approach to VLSI Circuit Simulation. C.A. Zukowski. ISBN 0-89838-176-2.
Multi-Level Simulation for VLSI Design. D.D. Hill and D.R. Coelho. ISBN 0-89838-184-3.
Relaxation Techniques for the Simulation of VLSI Circuits. J. White and A. Sangiovanni-Vincentelli. ISBN 0-89838-186-X.
VLSI CAD Tools and Applications. W. Fichtner and M. Morf, editors. ISBN 0-89838-193-2.
A VLSI Architecture for Concurrent Data Structures. W.J. Dally. ISBN 0-89838-235-1.
Yield Simulation for Integrated Circuits. D.M.H. Walker. ISBN 0-89838-244-0.
VLSI Specification, Verification and Synthesis. G. Birtwistle and P.A. Subrahmanyam. ISBN 0-89838-246-7.
Fundamentals of Computer-Aided Circuit Simulation. W.J. McCalla. ISBN 0-89838-248-3.
Serial Data Computation. S.G. Smith and P.B. Denyer. ISBN 0-89838-253-X.
Phonological Parsing in Speech Recognition. K.W. Church. ISBN 0-89838-250-5.
Simulated Annealing for VLSI Design. D.F. Wong, H.W. Leong, and C.L. Liu. ISBN 0-89838-256-4.
Polycrystalline Silicon for Integrated Circuit Applications. T. Kamins. ISBN 0-89838-259-9.
FET Modeling for Circuit Simulation. D. Divekar. ISBN 0-89838-264-5.
VLSI Placement and Global Routing Using Simulated Annealing. C. Sechen. ISBN 0-89838-281-5.
Adaptive Filters and Equalisers. B. Mulgrew and C.F.N. Cowan. ISBN 0-89838-285-8.
Computer-Aided Design and VLSI Device Development, Second Edition. K.M. Cham, S.-Y Oh, J.L. Moll, K. Lee, P. Vande Voorde and D. Chin. ISBN 0-89838-277-7.

AUTOMATIC SPEECH RECOGNITION

The Development of the SPHINX System

by

Kai-Fu Lee
Carnegie Mellon University

with a foreword by
Raj Reddy

KLUWER ACADEMIC PUBLISHERS
Boston/Dordrecht/London

Distributors for North America:
Kluwer Academic Publishers
101 Philip Drive
Assinippi Park
Norwell, Massachusetts 02061 USA

Distributors for the UK and Ireland:
Kluwer Academic Publishers
Falcon House, Queen Square
Lancaster LA1 1RN, UNITED KINGDOM

Distributors for all other countries:
Kluwer Academic Publishers Group
Distribution Centre
Post Office Box 322
3300 AH Dordrecht, THE NETHERLANDS

Library of Congress Cataloging-in-Publication Data

Lee, Kai-Fu
Automatic speech recognition : the development of the SPHINX system / by Kai-Fu Lee ; with foreword by Raj Reddy.
p. c.m. — (Kluwer international series in engineering and computer science ; 62. VLSI, computer architecture and digital signal processing)
Bibliography: p.
Includes index.
ISBN 0-89838-296-3
1. Automatic speech recognition. I. Title. II. Series: Kluwer international series in engineering and computer science. SECS 62. III. Series: Kluwer international series in engineering and computer science. VLSI, computer architecture, and digital signal processing.
TK7882.S65L44 1989 88-25930
006.4′54—dc19 CIP

Printed in the United States of America

Table of Contents

List of Figures

List of Tables

Foreword

Speech Recognition has a long history of being one of the difficult problems in Artificial Intelligence and Computer Science. As one goes from problem solving tasks such as puzzles and chess to perceptual tasks such as speech and vision, the problem characteristics change dramatically: knowledge poor to knowledge rich; low data rates to high data rates; slow response time (minutes to hours) to instantaneous response time. These characteristics taken together increase the computational complexity of the problem by several orders of magnitude. Further, speech provides a challenging task domain which embodies many of the requirements of intelligent behavior: operate in real time; exploit vast amounts of knowledge, tolerate errorful, unexpected unknown input; use symbols and abstractions; communicate in natural language and learn from the environment.

Voice input to computers offers a number of advantages. It provides a natural, fast, hands free, eyes free, location free input medium. However, there are many as yet unsolved problems that prevent routine use of speech as an input device by non-experts. These include cost, real time response, speaker independence, robustness to variations such as noise, microphone, speech rate and loudness, and the ability to handle non-grammatical speech. Satisfactory solutions to each of these problems can be expected within the next decade. Recognition of unrestricted spontaneous continuous speech appears unsolvable at present. However, by the addition of simple constraints, such as clarification dialog to resolve ambiguity, we believe it will be possible to develop systems capable of accepting very large vocabulary continuous speech dictation.

Work in speech recognition predates the invention of computers. However, serious work in speech recognition started in the late fifties with the availability of digital computers equipped with A/D converters. The problems of segmentation, classification, and pattern matching were explored in the sixties and a small vocabulary connected speech robot control task was demonstrated. In the early seventies, the role of syntax and semantics in connected speech recognition was explored and demonstrated as part of the speech understanding program. The seventies also witnessed the development of a number of basic techniques such as dynamic time warping, network representation, probabilistic modeling, beam search, and forward-backward algorithm. The early eighties witnessed a trend towards practical systems with very large vocabularies but computational and accuracy limitations made it necessary to require speaking one-word-at-a-time with a short pause between words.

With the start of the DARPA strategic computing program in speech, several new issues came to the fore: speaker independence; robustness, non-grammatical input and real time performance. The research reported by Kai-Fu Lee in this monograph represents an important milestone in one of these areas. For the first time it seems that high performance speaker independent recognition might be feasible without the tedium of having to train the system for each individual speaker. The research reported here builds on previous work of a number of researchers. It uses network representation, frame synchronous analysis and beam search techniques used in Dragon [Baker, 1975] and Harpy Systems [Lowerre, 1976]. The use of Cepstrum based signal representation was popularized by Furui [1986] and Shikano [1985]. It benefits from a number of techniques developed at IBM [Jelenik, 1980; Brown, 1987] for training of HMM such as deleted interpolation, tied transitions, null transitions, and the use of non-parametric models. Context dependent phone models were first used by IBM and BBN. However, a system based on many of these earlier techniques only resulted in mediocre performance (accuracy of 58%) in speaker independent recognition. Kai-Fu introduces a number of new techniques, which taken together lead to a speaker independent system performance (accuracy of 94%) comparable to the best speaker *dependent* systems. The availability of a large amount of multi-speaker training data, the use of automated learning techniques and the availability processors an order of magnitude faster than the prior generation have also been essential to achieving the necessary insights leading to a new milestone in the history of speech research.

Raj Reddy

Acknowledgements

This work evolved from my thesis research at Carnegie Mellon University. First and foremost, I would like to thank Raj Reddy, my thesis advisor, for his support, encouragement, and guidance. His thorough scientific approach and unending quest for excellence have been inspirational in the years of my thesis research.

I am most grateful to Peter Brown, who introduced me to hidden Markov models, patiently answered countless questions, and helped to shape and reshape many aspects of this work. It was reassuring to know that he was always there to help. I am also deeply indebted to Richard Schwartz, who has shared with me his valuable insights as well as details of the BBN system.

I also owe a great deal to the members of the CMU Speech Group. Hsiao-Wuen Hon has closely worked with me from the beginning, and is responsible for implementing various parts of SPHINX; without his help, I could not possibly be finished with this work by now. I am grateful to Roberto Bisiani, Alex Waibel, Richard Stern, Mei-Yuh Hwang, Brian Yamauchi, Joe Polifroni, Fil Alleva, Ron Cole, and Bob Weide for their help, as well as the rest of the Speech Group for their tolerance of my use of disk space (measured in gigabytes) and CPU time (measured in CPU-decades).

I thank Nicole Vecchi, Rachel Levine, and Roy Taylor for proofreading drafts of this book. I would also like to thank National Science Foundation and the Defense Advanced Research Projects Agency for supporting this work.

Finally, I wish to thank my family and friends for their support. I am especially grateful to my mother for everything she taught me and for all the sacrifices she made in my upbringing. Most of all, I thank my wonderful wife, Shen-Ling. She patiently looked after me when I was busy. She never had a word of complaint when I was negligent. She comforted me when I was discouraged. I may never be able to repay her, but fortunately, I have a lifetime to try.

Kai-Fu Lee

1 Introduction

In the past fifteen years, many speech recognition strategies have been proposed and implemented. These strategies span many sciences, including signal processing, pattern recognition, artificial intelligence, statistics, information theory, probability theory, computer algorithms, psychology, linguistics, and even biology. While a few systems demonstrated the feasibility of accurately recognizing human speech, they performed well because they imposed one or more of the following constraints: (1) speaker dependence, (2) isolated words, (3) small vocabulary, and (4) constrained grammar. We believe that the ultimate speech recognition system must be free of these constraints, and that the dependence on these constraints is caused by four deficiencies:

1. Lack of a sophisticated yet tractable model of speech.
2. Inadequate use of human knowledge of acoustics, phonetics, and lexical access in the recognizer.
3. Lack of consistent units of speech that are trainable and relatively insensitive to context.
4. Inability to account for between-speaker differences and speaker-specific characteristics.

In this monograph, we describe the implementation of a large-vocabulary speaker-independent continuous speech recognition system, SPHINX, that addresses these problems.

1.1. Constrained Speech Recognition: Achievements and Limitations

Considerable progress has been made in speech recognition in the past fifteen years. A number of successful systems have emerged. The most notable achievements include:

- Itakura [Itakura 75] of NTT introduced the dynamic time warp (DTW) for nonlinear alignment of speech. His system was tested on a single speaker, and achieved a 97.3% accuracy on a 200-word isolated word task without use of grammar.
- The DRAGON System of CMU [Baker 75a] used uniform stochastic modeling for all knowledge sources. On a 194-word speaker-dependent continuous task, DRAGON recognized 84% of the words correctly.[1]
- The HEARSAY System of CMU [Lesser 75] utilized a *blackboard* structure through which all knowledge sources communicate. It recognized 87% of the words correctly from a 1011-word speaker-dependent continuous task with a limiting syntax. The perplexity[2] of this task was 4.5.
- The HARPY System of CMU [Lowerre 76] combined the advantages of DRAGON and HEARSAY; it used network representation and beam search to improve the efficiency of the search. It achieved a percent correct rate of 97% on the same 1011-word task used by HEARSAY.
- Bell Labs [Wilpon 82] used clustering techniques to create robust templates for speaker-independent isolated-word recognition. Wilpon *et al.* reported an accuracy of 91% on a 129-word isolated-word speaker-independent task.
- The FEATURE System of CMU [Cole 83] used a feature-based approach that led to a speaker-independent system capable of recognizing isolated English letters with an accuracy over 90%, without grammar.
- The Tangora System [IBM 85] of IBM was the first to attempt a natural very-large-vocabulary task. It achieved over 97%

---

[1] When evaluating continuous speech recognizers, two measures have been used: *word accuracy* and *percent correct*. *Word accuracy* is lower than *percent correct* by the *insertion rate*. See Appendix Section I.2 for a detailed explanation.

[2] Perplexity is an information theoretic measure of a task's difficulty. It is roughly the number of choices at each decision point. See Appendix I.1 for the definition.

accuracy for speaker-dependent recognition of isolated word sentences using a 5000-word vocabulary with a natural-language-like grammar with perplexity 160. It was also capable of recognizing continuous speech, but with degraded and slower performance.

- The BYBLOS System of BBN [Chow 87, Kubala 88] successfully applied context-dependent modeling of phonemes. Although BYBLOS was speaker-dependent, it had an adaptive mode in which relatively short enrollment time is adequate for good performance. It achieved a 93% recognition rate on a 997-word continuous task with a much looser grammar (perplexity 60) than HARPY.
- Bell Labs [Rabiner 88a] produced the highest accuracy to date on speaker-independent connected digit recognition. Using continuous HMM technology with multiple models and mixtures, a sentence recognition rate of 97.1% was reached without use of grammar.

Each of the above systems attained very impressive accuracy, and have some practical applications. Moreover, significant progress has been made toward relaxing all four constraints. Nevertheless, no system has been able to perform satisfactorily without imposing considerable constraints. In particular, *speaker independence* has been viewed as the most difficult constraint, and speaker-independent systems have been severely constrained in other dimensions. Table 1-1 illustrates this point. In subsequent sections, we will explain why these constraints are hard to overcome. We will also argue that these constraints are undesirable for the ultimate speech recognizer.

1.1.1. Speaker Independence

A speaker-independent system is capable of recognizing speech from any new speaker. Speaker independence has been viewed as the most difficult constraint to overcome. This is because most parametric representations of speech are highly speaker dependent, and a set of reference patterns suitable for one speaker may perform poorly for another speaker.

There are three approaches to speaker independence. The first approach is to use knowledge engineering techniques to find perceptually motivated speech parameters that are relatively invariant between speakers. The justification for this approach is that if an expert spectrogram reader can

	Speaker Independence	Continuous Speech	Large Vocabulary	Natural Task
NTT	No	No	No	No
DRAGON	No	Yes	No	No
HEARSAY	No	Yes	Yes	No
HARPY	No	Yes	Yes	No
BELL '82	Yes	No	No	No
FEATURE	Yes	No	No	No
TANGORA	No	No	Yes	Yes
BYBLOS	No	Yes	Yes	No
BELL '88	Yes	Yes	No	No

Table 1-1: Important speech recognition systems and their freedom from constraints.

read spectrograms with high accuracy [Cole 80], it should be possible to find the invariant parameters such an expert employs. Furthermore, if these invariant parameters can be found, then speaker-independent recognition is as easy as speaker-dependent recognition. This idea has been carried out by many researchers [Haton 84, Zue 85, Thompson 87, Cole 86a]. In particular, Cole *et al.* achieved high accuracy on very limited tasks [Cole 83], but have not been successful on more difficult ones [Cole 86b].

The second approach is to use multiple representations for each reference to capture the between-speaker variations. The most well-known studies were performed by researchers at Bell Laboratories on clustering [Levinson 79, Rabiner 79, Rabiner 81]. Typically, each word in the vocabulary is uttered by many speakers; these multiple examples are then divided into several clusters, and a prototype is generated from each cluster. Lee [Lee 86] introduced a learning algorithm that clusters and generalizes at a subword level. Like the knowledge engineering approach, the multi-representation approach produced good results for limited tasks, but has not been successfully extended to a large-vocabulary task. Also, the recognizer knows the various characteristics of a speaker after a sentence or two. However, that information is not used and all the multiple examples from all the speakers are continually used—a slow process that can introduce otherwise avoidable errors.

The final category tries to use this knowledge about the speaker by adapting the recognizer to a new speaker. Speaker adaptation begins with an existing set of parameters, and a small number of adaptation sentences from the new speaker. These sentences are used to modify the parameters so that they are adjusted to the new speaker. Studies that tried to achieve speaker independence through adaptation include [Stern 83, Brown 83, Shikano 86a, Schwartz 87, Feng 88]. Strictly speaking, however, these systems are not truly speaker independent.

In spite of these efforts, no recognizer has achieved a reasonable accuracy on a large speaker-independent task. The successful speaker-independent systems in Table 1-1 were tested on very small vocabularies. To illustrate the relative difficulty of speaker-independent recognition, Levinson *et al.* [Levinson 77] reported that recognition accuracy degraded from 88.3% (98.2% for top 5 choices) for speaker-dependent to 65.1% (80% for top 5) for speaker-independent. Lowerre [Lowerre 77] reported that a speaker-dependent recognition rate of 98% degrades to 93% in speaker-independent mode. Finally, Shikano *et al.* [Shikano 86a] reported 48% accuracy for using one speaker's templates to recognize another speaker's speech, while 90% accuracy is possible for speaker-dependent recognition. Many researchers have used a rule of thumb that stated: for the same task, speaker-independent systems will have three to five times the error rate of speaker-dependent ones.

Because of these difficulties, most speech recognition systems are speaker dependent. In other words, they require a speaker to "train" the system before reasonable performance can be expected. This training phase typically requires several hundred sentences. Speaker-trained systems are useful for some applications; however, they have many problems:

1. The training session is an inconvenience to the user.
2. A large amount of processing is required before the system can be used. Typically, the user must wait many hours after speaking the training sentences.
3. Certain applications, such as telephone directory assistance or banking inquiries, cannot tolerate the delay of a training session.
4. Certain applications, such as courtroom dictation, may involve multiple speakers.
5. Considerable additional storage is needed if each speaker's

parameters are to be stored separately.

6. A speaker's voice may change over time due to stress, fatigue, sickness, or variations in microphone positioning.

Training a recognizer on other speakers' speech will lead to degraded results. However, we should not overlook an advantage of speaker-independent training, namely, much more data can be acquired to train speaker-independent systems. We will show in this monograph that this additional training, if properly used, can compensate for the less appropriate training material.

1.1.2. Continuous Speech

Continuous speech recognition is significantly more difficult than isolated word recognition. Its complexity is a result of three properties of continuous speech. First, word boundaries are unclear in continuous speech. In isolated-word recognition, word boundaries are known, and can be used to improve the accuracy and limit the search. In continuous speech, however, word boundaries are usually difficult to find. For example, in the phrase *this ship*, the /s/ of *this* is often omitted. In *we were away a year*, the whole sentence is one long vocalic segment, and word boundaries are difficult to locate. Second, *co-articulatory effects* are much stronger in continuous speech. Although we may try to pronounce words as concatenated sequences of phones, our articulators cannot move instantaneously to produce unaffected phones. As a result, a phone is strongly influenced by the previous and the following phones. In continuous speech, this effect occurs between words, and is much harder to predict. Moreover, as the speaking rate increases, within-word co-articulation is also accentuated. Third, *content words* (nouns, verbs, adjectives, etc.) are often emphasized, while *function words* (articles, prepositions, pronouns, short verbs, etc.) are poorly articulated. In particular, the phones in function words are often shortened, skipped, or distorted.

As a result, error rates increase drastically from isolated-word to continuous speech. Bahl *et al.* [Bahl 81a] reported error rates of 3.1% and 8.7% for isolated-word and continuous speech recognition, respectively, for a constrained 1000-word vocabulary. Moreover, the processing time for continuous speech was three times that of isolated-word speech.

In spite of these problems and degradations, we believe that it is important to work on continuous speech research, rather than be content with isolated-word recognition. While isolated-word recognition systems can have some applications, they are awkward for larger realistic tasks.

The primary advantage of automatic speech recognition is speed, since speech is the highest capacity human output channel. Talking is much faster than typing, currently the most popular form of data input. However, with isolated-word speech, this advantage diminishes substantially. To confirm this point, we computed statistics for two databases. The first database was recorded at IBM, and contained 1100 sentences spoken with pauses between words. The second database was recorded at Texas Instruments, and contained 1529 sentences spoken continuously. We computed the average word duration as the total amount of time divided by the total number of words in these 1529 sentence. The results are shown in Table 1-2. Also shown are the standard deviation and the number of words spoken per minute in either mode. It is possible to speak 170.45 words per minute in continuous speech—a great improvement over typing. On the other hand, using isolated-word input, a speaker can only dictate 71.6 words per minute, which is no better than the average typist.

	Isolated Words	Continuous Speech
Average Word Duration	0.838	0.352
Standard Deviation	0.171	0.082
Words Per Minute	71.60	170.45

Table 1-2: The average word duration (in seconds), standard deviation, and words per minute for isolated and continuous speech.

Another advantage of continuous speech is that it is a natural mode of human communication. Forcing pauses between words introduces artificiality, and reduces user-friendliness. Compared to isolated-word speech, the use of continuous speech input is much more likely to reduce the computer-phobia that is prevalent today. Finally, the unnaturalness of isolated-word speech may break the train of thought, when the user "thinks faster than he talks."

1.1.3. Large Vocabulary

Large vocabulary typically means a vocabulary of about 1000 words or more. Although vocabulary size is not the best measure of a task's difficulty, a number of problems arise when the vocabulary size is increased. The first problem is the inherent confusability of large vocabularies; researchers have found that the number of confusable words grows substantially when the vocabulary size reaches about 1000 words [Waibel 86].

With small vocabularies, each word can be modeled individually, because it is reasonable to expect sufficient training for a handful of words. It is also possible to store the parameters of each word model separately. However, as the vocabulary size increases, it is no longer possible to train each word explicitly, because neither the training nor the storage is available. Instead, some *subword unit* must be identified and used. Subword units usually lead to degraded performance because they cannot capture co-articulatory (inter-unit) effects as well as word models can. Rosenberg [Rosenberg 83] reported a two fold increase in error rate when word templates were replaced with demisyllable units. Paul and Martin [Paul 88] reported about a tenfold increase in error rate from word models to phone models.

Another difficulty is the complexity of search. For small vocabularies, it is possible to perform optimal searches; however, for large vocabularies, pruning will be necessary. Pruning may introduce search errors, which hurt recognition accuracy.

1.1.4. Natural Task

The final constraint is concerned with the grammar used by the speech recognition system. The difficulty of a grammar, or the amount of constraint imposed by a grammar, can be measured by *perplexity*, an information theoretic measurement of the average uncertainty at each decision point. Appendix I.1 contains more details on computing perplexity. While perplexity is not a perfect measure of grammar difficulty, it is the best standard measure available.

In most early systems [Lesser 75, Lowerre 76], the best results were obtained using grammars that were finite state networks of allowable sentences. These grammars had a perplexity of less than 5. Bahl *et al.* [Bahl

83a] showed that error rate can be extremely low with such grammars. They reported a percent incorrect rate of 0.1% for a grammar with perplexity 4.5, and 8.9% for a grammar with perplexity 24. Kimball, *et al.* [Kimball 86] reported that error rate increased from 1.6% to 4.5% when perplexity increased from 19 to 58. Thus, increasing perplexity usually results in a substantial loss of accuracy.

Although recent systems began to use more difficult grammars, the only speech recognition system capable of accepting natural-language-like input is IBM's TANGORA, which has a vocabulary of 20,000 words, and a trigram grammar[3] with perplexity over 200. This grammar was computed from 250 million words of training text. While the trigram grammar worked well for IBM, other researchers do not have the resources to train such grammars; therefore, progress in this direction will probably be the slowest.

Accepting grammars that have high perplexity is an important goal for the ultimate speech recognizer, because only such grammars can be highly versatile. For tasks such as dictation, it is simply not possible to construct lower perplexity grammars.

1.2. Relaxing the Constraints: The SPHINX System

We have briefly described several successful speech recognition systems, and showed that their impressive performances are aided by imposing several constraints. Moreover, we have explained why these constraints are undesirable in natural man-machine communications.

In this monograph, we investigate ways to overcome the small vocabulary, speaker dependence, and isolated word constraints. Although we cannot train a natural language recognizer with our limited training text and vocabulary, we will simulate the effect of higher perplexity by using looser grammars and presenting comparative results.

Overcoming each of these three constraints is extremely difficult. Error rate more than tripled [Levinson 77, Bahl 81a] when either the speaker dependence or the isolated word constraint was relaxed. A larger vocabulary

---

[3]A trigram grammar estimates the probability of a word in a sentence given the two previous words.

could also cause a substantial degradation in accuracy. Therefore, in order to overcome these constraints, we must address a number of fundamental problems before reasonable performance can be expected:

1. ***Poor modeling of speech.*** It is desirable to use a large number of speech parameters (or features) for speaker-independent recognition. Thus, we need a modeling technique that can account for many parameters. Due to the complexities introduced by freeing these constraints and the greater amount of training data available for speaker-independent recognition, efficient and automatic algorithms must exist for training and recognizing with our model. An appropriate modeling technique is also needed so that a large vocabulary does not create practical problems.

2. ***Lack of human knowledge.*** At the parametric level, it is useful to incorporate additional parameters that are relatively speaker independent. At the phonetic and lexical levels, it would also be helpful to incorporate human knowledge into the recognizer.

3. ***Lack of a good unit.*** To deal with a large vocabulary, we must use subword units. However, in continuous speech, adjacent phonemes may have strong effects on each other, even across word boundaries. The unit of speech must model this *co-articulatory* effect. Moreover, phonemes in function words are particularly distorted, and units that can account for this distortion would be useful.

4. ***Lack of learning and adaptation.*** Since acoustic properties are often not speaker independent, it is useful to learn characteristics of a speaker and adapt accordingly. However, the adaptation process must be rapid and non-intrusive—otherwise it degenerates into a speaker-dependent system.

We now describe SPHINX—an accurate, large-vocabulary, speaker-independent, continuous speech recognition system. Our implementation of SPHINX was guided by our conviction that the four fundamental problems must be remedied.

1.2.1. Hidden Markov Models: A Representation of Speech

Hidden Markov modeling (HMM) is a powerful technique capable of robust modeling of speech. An HMM is a parametric model that is particularly suitable for describing speech events. HMMs have two stochastic processes which enable the modeling not only of acoustic phenomena, but also of timescale distortions. Furthermore, efficient algorithms exist for accurate estimation of HMM parameters. Unlike other non-parametric and *ad-hoc* approaches, the forward-backward re-estimation algorithm for hidden Markov models is an instance of the EM algorithm [Baum 72]. As such, every iteration of the algorithm results in an improved set of model parameters. Hidden Markov models are a succinct representation of speech events; therefore, they require less storage than many other strategies.

Hidden Markov models were first used independently by Baker at CMU [Baker 75a, Baker 75b] and the IBM Speech Group [Bakis 76, Jelinek 76] in speech recognition. Recently, researchers have successfully applied these HMM techniques to model various units of speech, and to recognize speech under different constraints [IBM 85, Paul 86, Chow 86, Rabiner 88a].

We use hidden Markov models to represent all knowledge sources, from phones to words to sentences. In addition to applying the standard algorithms, we have also experimented with a number of HMM variations.

1.2.2. Adding Human Knowledge

Although HMMs have been used to represent units of speech in the past, they were usually speaker-dependent models. We extend HMMs to speaker-independent modeling by adding various types of human knowledge about speech.

Most speech recognition systems, including HMM-based ones, rely on relatively simple speech parameters, such as FFT or LPC coefficients. Many recent studies have revealed the usefulness of other parameters. Waibel [Waibel 86] showed that the use of prosodic parameters, such as duration, intensity, and stress, improved word recognition significantly. Furui [Furui 86] showed that the addition of differential coefficients and power reduced the error rate of a speaker-independent recognizer from 6.4% to 2.4%. Shikano [Shikano 86b] added similar coefficients and obtained significant improvements in phone recognition.

The above studies attempted to enhance traditional techniques with new parameters. A number of other researchers [Cole 83, Haton 84, Adams 86, Thompson 87], encouraged by Victor Zue's success in spectrogram reading [Cole 80], viewed speech recognition as a process of imitating expert spectrogram readers. By examining visual representations of the spectrogram, useful features can be extracted and used in recognition. Feature-based systems directly and extensively use these perceptually motivated parameters; however, they suffer from excessive dependence on human guidance, difficulty in integration with other knowledge sources, and lack of a trainable and tractable model. Consequently, they have had only limited success.

In view of these difficulties, we will only use knowledge engineering techniques to improve SPHINX within an HMM-framework. Both fixed-width features and variable-width features are added to SPHINX. Fixed-width features are those that can be measured for each fixed time frame. For example, cepstrum coefficients, power, or differenced coefficients are all examples of fixed-width features. Fixed-width features can be easily integrated in hidden Markov models by adding new parameters for each frame. We model these additional parameters with multiple vector quantized (VQ) codebooks. Variable-width features include phoneme duration, duration for aperiodic energy before and after vowels, frequency location of formants, and many others [Cole 83]. Since variable-width features are difficult to implement given the Markov assumption,[4] these features cannot be easily modeled in their original form. However, we will examine ways in which they can be integrated into an HMM-based recognizer without training.

Finally, there is a large body of knowledge about the phonology and phonologic variations of English. We use this knowledge to improve the set of phones, the word pronunciation dictionary, as well as phonological rules.

1.2.3. Finding a Good Unit of Speech

Given that we will use HMMs to model speech, we need to define the fundamental unit that an HMM will model. This unit of speech should be not only trainable and well-defined, but also relatively insensitive to context. The most obvious units are words and phonemes. Word models, while suitable for small-vocabulary recognition, are not a practical choice for large-

---

[4]The Markov assumptions stipulate that each fixed-width frame is dependent only on its state, and are conditionally independent of the past. See Section 2.1.

vocabulary recognition because of the great amount of storage and training data required. Although phonemes are a standard and well understood unit, and are easily trainable, they are highly dependent on left, right, and word-dependent contexts [Schwartz 85, Chow 86].

In order to capture co-articulatory (or contextual) effects, researchers have proposed the use of multi-phoneme units, such as demisyllables [Rosenberg 83], diphones [Schwartz 80, Klatt 86], and syllables [Hunt 80]. While the central portions of these units are unaffected by context, the beginning and ending portions are still susceptible to some contextual effects. Moreover, there are a large number (over 1000 demisyllables, 2500 diphones, and 20,000 syllables) of these units. This is undesirable for an HMM system because considerable training data is needed to estimate the parameters for each model. Finally, experiments [Rosenberg 83] showed that a demisyllable-based recognizer performed substantially worse than a word-based recognizer.

Bahl *et al.* [Bahl 80a] first proposed context-dependent modeling of phones (or triphone modeling). Schwartz *et al.* [Schwartz 84, Schwartz 85] at BBN successfully applied this idea by creating a phoneme model for each left/right phonemic context. However, this leads to a large number of poorly trained models. This problem was dealt with by interpolating (or averaging) the context-dependent models with context-independent ones. We will refer to phone models that take into account left and right contexts *triphone models.*

Another problem with triphone modeling is that triphone models do not account for the similarity between certain contexts. We believe that it is possible to generalize further by merging similar contexts. We use an information-theoretic measure to cluster similar contexts into *generalized triphone models.* This not only results in substantially fewer context-dependent units, but also provides more training data for each unit. Instead of using hand-tuned weights as BBN did [Schwartz 85], we use deleted interpolation [Jelinek 80] to combine detailed models with robust ones. Deleted interpolation is an elegant EM (Estimate-Maximize) algorithm that estimates the weight of the models based on how well each model predicts unseen data.

We introduce another novel unit, the *function-word-dependent phone model,* which explicitly models phones in function words. These units focus on the most confusable sub-vocabulary, and can be well-trained because function words occur frequently. *Function-word-dependent phone models* are also interpolated with *context-independent phone models* using deleted interpolation.

1.2.4. Speaker Learning and Adaptation

While the above representations and techniques will improve recognition, they do not account for vocal tract differences between different speakers, nor do they capture speaker-specific characteristics. We propose two algorithms that adapt the HMM system to the speaker.

The first algorithm is based on *speaker cluster selection* [Shikano 86a, Shikano 86b]. *Speaker cluster selection* assumes that speakers can be divided into clusters, within which the speakers are similar. It has been shown that inappropriate VQ codebooks are accountable for a major part of speaker differences [Shikano 86a, Schwartz 87]. Therefore, to adapt to a speaker, we could first identify the speaker cluster that best resembles his or her characteristics, and use the codebook for that speaker cluster for recognition. Cluster-specific HMM parameters can be derived by only training on speakers in the cluster, or by training on all speakers, and then converting to a cluster-specific codebook through a probabilistic mapping.

While the speaker clustering assumption is surely reasonable, limited training data permits only a small number of speaker clusters. With so few clusters, it is only possible to differentiate very obvious general differences, such as vocal tract length. There are many minor speaker differences, such as nasalization, flapping, and stop deletion, that cannot be taken into account by speaker clustering. Therefore, we introduce the *interpolated re-estimation* algorithm, which begins with a speaker-independent (or cluster-dependent) codebook and HMMs. The codebook remains unchanged, while the HMM parameters are modified to better suit the speaker. With a small number of adaptation sentences and the correct word sequences in these sentences, speaker-dependent parameters are derived by running the forward-backward algorithm. Then, we interpolate these parameters, which are appropriate but poorly trained, with the speaker-independent parameters, which are less appropriate but well-trained. In addition, we also derive additional statistics based on these sentences, including tied-phone-state parameters, co-occurrence smoothed parameters, and similar-speaker parameters. All five sets of parameters are linearly combined using weights that depend on the reliability of the speaker-dependent parameters. These weights are determined automatically using *deleted interpolation* [Jelinek 80].

Our approach to speaker adaptation begins with a speaker-independent system. Thus, there is no delay in enrolling the speaker, and the system can adapt to the input speaker in an incremental and non-intrusive manner.

1.3. Summary and Monograph Outline

The techniques described in the previous sections were implemented in SPHINX. SPHINX achieved speaker-independent word accuracies of 71%, 94%, and 96% on the 997-word DARPA resource management task with grammars of perplexity 997, 60, and 20. With speaker adaptation, error rates are further reduced by about 5–10%. These are the best results reported for similar systems. In fact, they are slightly better than the *speaker-dependent* results reported by BBN [Kubala 88].

By utilizing perceptual knowledge and stochastic modeling, by integrating knowledge and learning, and by fully utilizing abundant training, the SPHINX Speech Recognition System has bridged the gap between speaker-dependent and speaker-independent systems.

In this monograph, we will first provide background information on hidden Markov modeling of speech in Chapter 2. In Chapter 3, we will describe our task and the two databases we use for training.

Starting from Chapter 4, we will describe various implementations of SPHINX by incrementally introducing our solutions to the four fundamental problems. At the end of each chapter, we will present our results, and the best version will be used in the succeeding chapters. This approximates our progress on SPHINX over the past two years. We hope that this will illustrate the utility of each enhancement, as well as share with readers our excitement after each significant improvement.

In Chapter 4, we will describe the baseline version of SPHINX using standard speech parameters and HMM techniques. We also discuss some HMM improvements and tuning experiments. Chapter 5 considers three knowledge-based improvements to SPHINX, namely, the use of fixed-width parameters, variable-width parameters, and phonetic/lexical knowledge. Chapter 6 describes two novel units of speech: *function-word-dependent phone* models and *generalized triphone phone* models. Next, the two learning algorithms, *speaker cluster selection* and *interpolated re-estimation*, are described and evaluated in Chapter 7. Chapter 8 summarizes SPHINX results, compares them against similar systems, and analyzes the errors SPHINX made. Finally, Chapter 9 contains the conclusions of this work.

2 Hidden Markov Modeling of Speech

Hidden Markov models (HMM) were first described in the classic paper by Baum [Baum 72]. Shortly afterwards, they were extended to automatic speech recognition independently at CMU [Baker 75a] and IBM [Bakis 76, Jelinek 76]. It was only in the past few years, however, that HMMs became the predominant approach to speech recognition, superseding dynamic time warping.

In this chapter, we will first define hidden Markov models and present algorithms for evaluating, decoding, and learning with HMMs. Next, we will discuss how the speech recognition problem can be formulated as an HMM problem. Finally, we will look at some implementational issues for training HMMs and using HMMs in recognition. We will begin at a level that any reader with some background in probability can follow, and continue in sufficient detail so that if a reader wanted to implement an HMM recognizer, implementational difficulties could be minimized.

Interested readers are also directed to [Baker 75b, Jelinek 76, Bahl 83a, Levinson 83, Rabiner 86, Rabiner 88b] for further reading on the fundamentals of hidden Markov models.

2.1. Definition of a Hidden Markov Model

A hidden Markov model is a collection of states connected by transitions. Each transition carries two sets of probabilities: a transition probability, which provides the probability for taking this transition, and an output probability density function (pdf), which defines the conditional probability of emitting each output symbol from a finite alphabet given that a transition is taken.[5] Figure 2-1 shows an example of a hidden Markov model with two output symbols, A and B.

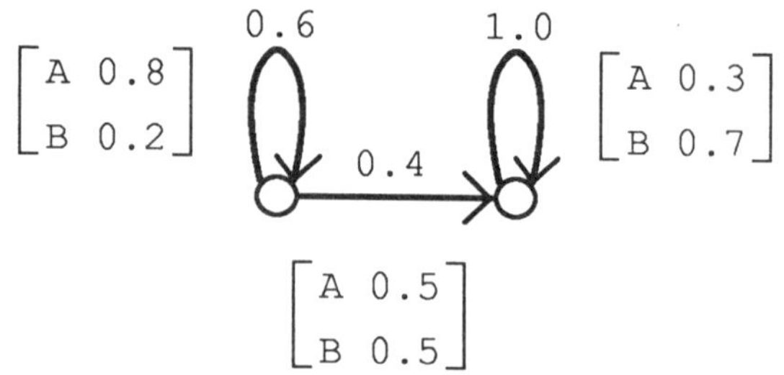

Figure 2-1: A simple hidden Markov model with two states, and two output symbols, A and B.

A hidden Markov model is defined by:

- $\{s\}$—A set of states including an initial state S_I and a final state S_F.[6]
- $\{a_{ij}\}$—A set of transitions where a_{ij} is the probability of taking a transition from state i to state j.
- $\{b_{ij}(k)\}$—The output probability matrix: the probability of emitting symbol k when taking a transition from state i to state j.

Since both a and b are probabilistic, they must satisfy the following properties:

$$a_{ij} \geq 0, \quad b_{ij}(k) \geq 0, \quad \forall i,j,k \tag{1}$$

$$\sum_j a_{ij} = 1 \ \forall i \tag{2}$$

$$\sum_k b_{ij}(k) = 1 \ \forall i,j \tag{3}$$

a and b can be written as:

---

[5]Most literature associates output distributions with states, not transitions. But that is just a special case of our definition.

[6]Actually, multiple initial and final states can be included easily (see Section 2.3.2), but we make this assumption to simplify notations.

$$a_{ij} = P(X_{t+1}=j \mid X_t=i) \tag{4}$$

$$b_{ij}(k) = P(Y_t=k \mid X_t=i, X_{t+1}=j) \tag{5}$$

where $X_t=j$ means the Markov chain was in state j at time t, and $Y_t=k$ means the output symbol at time t was k. We will use the random variable Y to represent the probabilistic function of a stationary Markov chain X. Both X and Y are generated by a hidden Markov model; however, Y, the output sequence, is directly observed, while X, the state sequence, is *hidden*.

In a first-order hidden Markov model, there are two assumptions. The first is the *Markov assumption*:

$$P(X_{t+1}=x_{t+1} \mid X_1^t=x_1^t) = P(X_{t+1}=x_{t+1} \mid X_t=x_t) \tag{6}$$

where X_i^j represents the state sequence $X_i, X_{i+1}, \ldots, X_j$. Equation 6 states that the probability that the Markov chain is in a particular state at time $t+1$ depends only on the state of the Markov chain at time t, and is conditionally independent of the past.

The second assumption is the *output-independence assumption*:

$$P(Y_t=y_t \mid Y_1^{t-1}=y_1^{t-1}, X_1^{t+1}=x_1^{t+1}) = P(Y_t=y_t \mid X_t=x_t, X_{t+1}=x_{t+1}) \tag{7}$$

where Y_i^j represents the output sequence $Y_i, Y_{i+1}, \ldots, Y_j$. Equation 7 states that the probability that a particular symbol will be emitted at time t depends only on the transition taken at that time (from state x_t to x_{t+1}), and is conditionally independent of the past.

Although these assumptions severely limit the memory of first-order hidden Markov models, they reduce the number of parameters. As we will see, they also make learning and decoding algorithms extremely efficient.

2.2. Three HMM Problems

Given the definition of hidden Markov models, there are three problems of interest:

- **The Evaluation Problem**—Given a model and a sequence of observations, what is the probability that the model generated the observations?

- **The Decoding Problem**—Given a model and a sequence of observations, what is the most likely state sequence in the model that produced the observations?
- **The Learning Problem**—Given a model and a set of observations, what should the model's parameters be so that it has a high probability of generating the observations?

If we could solve the *evaluation* problem, we would have a way of scoring the match between a model and an observation sequence, which could be used for isolated-word recognition. If we could solve the *decoding* problem, we could find the best matching state sequence given an observation sequence, which could be used for continuous speech recognition. Most importantly, if we could solve the *learning* problem, we would have the means to automatically learn the parameters given an ensemble of training data. In this section, we will uncover solutions to these three problems.

2.2.1. The Evaluation Problem : The Forward Algorithm

The evaluation problem can be stated as: given a model, **M**, with parameters $\{s\}, \{a\}, \{b\}$, compute the probability that it will generate a sequence y_1^T. This involves summing the probabilities of all paths of length T:

$$P(Y_1^T = y_1^T) = \sum_{x_1^{T+1}} P(X_1^{T+1} = x_1^{T+1}) P(Y_1^T = y_1^T \mid X_1^{T+1} = x_1^{T+1}) \tag{8}$$

In other words, to compute the probability of the sequence y_1^T, we enumerate all paths x_1^{T+1} of length T that generate y_1^T, and sum all their probabilities. The probability of each path x_1^{T+1} is the product of the transition probabilities and the output probabilities of each step in the path.

The first factor (transition probability) in Equation 8 can be re-written by applying the Markov assumption:

$$P(X_1^{T+1} = x_1^{T+1}) = \prod_{t=1}^{T} P(X_{t+1} = x_{t+1} \mid X_t = x_t) \tag{9}$$

The second factor (output probability) in Equation 8 can be re-written by applying the output-independence assumption:

$$P(Y_1^T = y_1^T \mid X_1^{T+1} = x_1^{T+1}) = \prod_{t=1}^{T} P(Y_t = y_t \mid X_t = x_t, X_{t+1} = x_{t+1}) \tag{10}$$

Substituting Equation 9 and 10 into 8, we have:

$$P(Y_1^T = y_1^T) = \sum_{x_1^{T+1}} \prod_{t=1}^{T} P(X_{t+1} = x_{t+1} \mid X_t = x_t) P(Y_t = y_t \mid X_t = x_t, X_{t+1} = x_{t+1}) \tag{11}$$

We can directly evaluate Equation 11 from the HMM parameters a and b. However, the evaluation of Equation 11 requires enumeration of all paths with length T, which is clearly exponential.

Fortunately, because of our assumptions, the probability of each quantity $P(X_{t+1} = x_{t+1} \mid X_t = x_t) P(Y_t = y_t \mid X_t = x_t, X_{t+1} = x_{t+1})$ only involves y_t, x_t, and x_{t+1}. It is, therefore, possible to compute $P(Y_1^T = y_1^T)$ with recursion on t. Let's define:

$$\alpha_i(t) = \begin{cases} 0 & t = 0 \wedge i \neq S_I \\ 1 & t = 0 \wedge i = S_I \\ \sum_j \alpha_j(t-1)\, a_{ji}\, b_{ji}(y_t) & t > 0 \end{cases} \tag{12}$$

$\alpha_i(t)$ is the probability that the Markov process is in state i having generated y_1^t. Clearly then,

$$P(Y_1^T = y_1^T) = \alpha_{S_F}(T) \tag{13}$$

Figure 2-2 illustrates the computation of α as a sweep through a *trellis* using the HMM in Figure 2-1 and the observation sequence `A A B`. Each cell indicates the cumulative probability at a particular state (row) and time (column). An arrow in Figure 2-2 indicates that a transition from its origin state to its destination state is legal. Consequently, arrows are only allowed between adjacent columns. As Equation 12 indicates, the computation begins by assigning 1.0 to the initial state and 0.0 to all other states at time 0. The other cells are computed *time-synchronously* from left to right. Each column of states for time t is completely computed before going to time $t+1$, the next column. When the states in the last column have been swept, the final state in the final column contains the probability of generating the observation sequence.

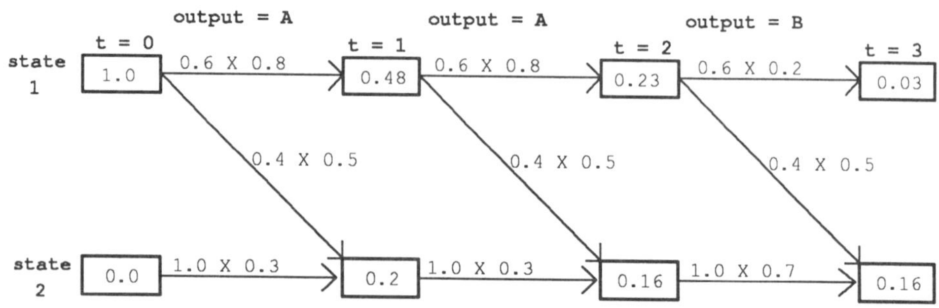

Figure 2-2: A trellis in the forward computation.

This algorithm is called the *forward pass* or the *forward algorithm.* It enables us to evaluate the probability that an observation sequence was generated by an HMM, **M**, or $P(y \mid \mathbf{M})$. However, in speech recognition, we need to find $P(\mathbf{M} \mid y)$. By Bayes rule, we have

$$P(\mathbf{M} \mid y) = \frac{P(y \mid \mathbf{M})\, P(\mathbf{M})}{P(y)} \tag{14}$$

Since $P(y)$ is constant for a given input, the task of recognizing an observation sequence involves finding the model that maximizes $P(y \mid \mathbf{M}) \cdot P(\mathbf{M})$. $P(y \mid \mathbf{M})$ can be evaluated by the forward algorithm, and $P(\mathbf{M})$ is a probability assigned by the *language model.* In the case of a language model where all words are equally likely (the $P(m)$'s are equal for all m), only the first factor is needed.

2.2.2. The Decoding Problem : The Viterbi Algorithm

While the forward algorithm computes the probability that an HMM generated an observation sequence, it does not provide a state sequence. In many applications, it may be desirable to have such a sequence. As we will see later, finding the optimal state sequence could be used for segmentation and recognition of speech.

Unfortunately, by definition, the state sequence is *hidden* in an HMM. The best we can do is to produce the *state sequence that has the highest probability of being taken while generating the observation sequence.* To do that, we need only modify the forward pass slightly. In the forward pass, we summed probabilities that came together. Now, we need to choose and remember the maximum.

$$v_i(t) = \begin{cases} 0 & t = 0 \wedge i \neq S_I \\ 1 & t = 0 \wedge i = S_I \\ \mathbf{MAX}_j v_j(t-1)\, a_{ji} b_{ji}(y_t) & t > 0 \end{cases} \tag{15}$$

To uncover the most likely state sequence, we must remember the best path to each cell, which is the concatenation of the best path to its predecessor state and the best step to the cell.

This algorithm is known as the *Viterbi algorithm* [Viterbi 67]. It can be used for segmentation, annotation, and recognition. In Section 2.4.3.2 we will discuss its merits and shortcomings when applied to speech recognition.

2.2.3. The Learning Problem : The Forward-Backward Algorithm

The learning problem involves optimizing HMM parameters given an ensemble of training data. It is the most difficult of the three problems, because there is no known analytical method to solve for the parameters in a maximum likelihood model. Instead, an iterative procedure or a gradient descent technique must be used. Here we will describe an iterative procedure, the *forward-backward algorithm*, also known as the *Baum-Welch algorithm.*

In Equation 12 of Section 2.2.1, we defined $\alpha_i(t)$, or the probability that an HMM **M** has generated y_1^t and is in state i. We now define its counterpart, $\beta_i(t)$, or the probability that **M** is in state i, and will generate y_{t+1}^T. Like α, β can be computed with recursion on t:

$$\beta_i(t) = \begin{cases} 0 & i \neq S_F \wedge t = T \\ 1 & i = S_F \wedge t = T \\ \sum_j a_{ij} b_{ij}(y_{t+1})\, \beta_j(t+1) & 0 \leq t < T \end{cases} \tag{16}$$

Let us now define $\gamma_{ij}(t)$, which is the probability of taking the transition observation sequence y_1^T:

$$\gamma_{ij}(t) = P(X_t = i, X_{t+1} = j \mid y_1^T)$$
$$= \frac{\alpha_i(t-1)\, a_{ij}\, b_{ij}(y_t)\, \beta_j(t)}{\alpha_{S_F}(T)} \tag{17}$$

$\alpha_{S_F}(T)$, also known as the *alpha terminal*, is the probability that **M** generated y_1^T. Now, the expected number (or count) of transitions from state i to j given y_1^T at any time is simply $\sum_{t=1}^{T} \gamma_{ij}(t)$, and the expected number of counts from state i to any state at any time is $\sum_{t=1}^{T} \sum_k \gamma_{ik}(t)$. Then, given some initial parameters, we could recompute the probability of taking the transition from state i to state j as:

$$\bar{a}_{ij} = \frac{\sum_{t=1}^{T} \gamma_{ij}(t)}{\sum_{t=1}^{T} \sum_{k} \gamma_{ik}(t)} \tag{18}$$

Similarly, $b_{ij}(k)$ can be re-estimated as the ratio between the frequency that symbol k is emitted and the frequency that any symbol is emitted, or:

$$\bar{b}_{ij}(k) = \frac{\sum_{t:y_t = k} \gamma_{ij}(t)}{\sum_{t=1}^{T} \gamma_{ij}(t)} \tag{19}$$

Equations 18 and 19 are both instances of the Baum-Welch algorithm [Baum 72]. As such, every re-estimate is guaranteed to increase $\alpha_{S_F}(T)$, unless a critical point is already reached, in which case the re-estimate will remain the same. Several proofs of this have been presented in the literature [Baum 72, Meilijson 87]. We will sketch Baum's proof here [Baum 72, Levinson 83].

First, let's define:

$$u_i = P(i, y \mid \mathbf{M}) \tag{20}$$

$$v_i = P(i, y \mid \overline{\mathbf{M}}) \tag{21}$$

where i is a state sequence, and $\mathbf{M}$ and $\overline{\mathbf{M}}$ are two HMMs. Thus, by summing over all i's:

$$\sum_i u_i = P(y \mid \mathbf{M}) \tag{22}$$

$$\sum_i v_i = P(y \mid \overline{\mathbf{M}}) \tag{23}$$

It can be shown through Jensen's inequality [Baum 72] that

$$\ln \frac{P(y \mid \overline{\mathbf{M}})}{P(y \mid \mathbf{M})} \geq \frac{1}{P(y \mid \mathbf{M})}[Q(\mathbf{M},\overline{\mathbf{M}}) - Q(\mathbf{M},\mathbf{M})] \tag{24}$$

where $Q(\mathbf{M},\overline{\mathbf{M}}) = \sum_i u_i \ln v_i$

This suggests that if a model $\overline{\mathbf{M}}$ can be derived from $\mathbf{M}$ such that the right hand side of Equation 24 is positive, then we have a way of improving $\mathbf{M}$, which can be accomplished by maximizing $Q(\mathbf{M},\overline{\mathbf{M}})$.

By decomposing $Q(\mathbf{M},\overline{\mathbf{M}})$ and using the fact that the function

$$F(x) = \sum_i c_i \ln x_i, \tag{25}$$

subject to the constraint $\sum_i x_i = 1$, attains its unique global value when

$$x_i = \frac{c_i}{\sum_{i'} c_{i'}}, \tag{26}$$

it can be shown [Baum 72] that Equations 18 and 19 will always result in an improvement to $\alpha_{S_F}(T)$, unless it is already at a critical point.

The above proof sketch clearly leads to the following iterative algorithm, which is known as the forward-backward algorithm, or Baum-Welch Algorithm:

1. Guess an initial set of parameters $\{a, b\}$.
2. Compute $\bar{a}$ and $\bar{b}$ according to the re-estimation formulas in Equations 18 and 19.
3. Set a to $\bar{a}$ and b to $\bar{b}$.
4. If some convergence criteria are not met, go to step 2.

This type of iterative algorithms is known as EM (Estimate-Maximize) Algorithm.

2.3. Implementational Issues

2.3.1. Tied Transition

In our previous discussion, we have assumed that each transition has a separate output probability density function (pdf), or b. In practice, this may be undesirable. For example, if we wanted to train a word model with 10 sequential states and 20 transitions, each with a distinct output pdf, we would have to estimate a tremendous number of parameters. Instead, we could use a lot of states to model duration, and allow adjacent sets of transitions to share the same output pdf. In another example, since certain states in a model may be the same (for example, the /s/'s and /ih/'s in *Mississippii*), we want different states to share the same output pdf's.

We can allow different states to share the same output pdf by a simple modification of the re-estimation formula. For transitions, let us define $\tau(i,j)$ as the set of transitions that the transition $(i \rightarrow j)$ is tied to, and $\sigma(i)$ as the set of states that state i is tied to. Then $\bar{a}_{ij}$ can be re-estimated as follows:

$$\bar{a}_{ij} = \frac{\sum_{i',j' \in \tau(ij)} \sum_{t=1}^{T} \gamma_{i'j'}(t)}{\sum_{i' \in \sigma(i)} \sum_{t=1}^{T} \sum_{k} \gamma_{i'k}(t)} \tag{27}$$

Similarly $\bar{b}_{ij}(k)$ can be re-estimated as below:

$$\overline{b}_{ij}(k) = \frac{\sum_{i',j' \in \tau(ij)} \sum_{t:y_t = k} \gamma_{i'j'}(t)}{\sum_{i',j' \in \tau(ij)} \sum_{t=1}^{T} \gamma_{i'j'}(t)} \tag{28}$$

The maximum likelihood property of the forward-backward algorithm still holds with tied transitions [Jelinek 80]. As a side note, if we hold any set of probabilities in an HMM fixed, maximum likelihood estimates of the remaining parameters can still be estimated the same way as before.

2.3.2. Null Transitions

Null transitions are a convenient notation that could be used to reduce the total number of states. A null transition, like a normal transition, has a transition probability. However, a null transition does not emit an output symbol, and therefore does not consume a unit of time. In other words, if there is a null transition from state a to state b, a Markov process in state a can switch to state b without using any time or emitting any symbols.

We need to modify all our algorithms to take null transitions into account. A special case for null transitions has to be included in the α and β computations because no output symbols are emitted. Jelinek and Mercer [Jelinek 80] show that the introduction of null transitions does not cause any problems for maximum likelihood estimation.

To simplify notations, we have previously assumed a single initial and final state. Actually, multiple initial and final states can be included trivially by using null transitions. We would add two additional states, representing the new initial and final states, and add a null transition from the new initial state to each of the old initial states, and from each of the old final states to the new final state. A more parsimonious solution is to include an initial state probability, and to sum over all final states [Rabiner 86, Brown 87].

2.3.3. Initialization

One issue we have not addressed is how the statistics are initialized for the forward-backward training. While the forward-backward algorithm guarantees an improvement every iteration, it does not guarantee finding a global maximum. If the initialization is poor, a poor local maximum may be found. In particular, if a probability is initialized to be zero, it will remain zero with every iteration.

One initialization technique is to use hand-marked training data, and initialize the statistics accordingly [Schwartz 85, Lee 87]. With phonetic models, all output pdf's of the phone can be initialized identically. However, with word models, different states of the same word model may have drastically different statistics, and a more sophisticated initialization algorithm is needed. Juang *et al.* [Juang 85a] proposed a segmental K-means procedure to find more accurate initializations.

Our experience indicates that complex initialization algorithms are not necessary for discrete density pdf's. A simple uniform distribution is likely to be sufficient, assuming a reasonable amount of training data exists. If the parameters are allowed too many degrees of freedom (either too many parameters or continuous parameters) with respect to the amount of training, more sophisticated forms of initialization are needed. This is corroborated by Paul and Martin [Paul 88] who found that bootstrapping (or initialization) became important when the number of parameters increased.

2.3.4. Scaling or Log Compression

Due to the inaccuracies of the Markov assumption and the output independence assumption, the probability of a word (or sentence) will approach zero during forward computation. After hundreds of frames, this probability would underflow the floating point representation of almost any digital computer. As a result, some type of re-scaling is necessary. The most popular method of scaling is to divide all probabilities by the sum of all the α's in a column after that column has been processed in the forward pass. This sum, the *scaling factor*, has to be saved for each rescaled column, so that at least the log probability can be computed at the end of the forward pass. The same scaling factors are used in the backward pass. Recall that re-estimation involves division of one γ by the sum of many γ's. Since both the numerator and the denominator had been scaled identically, the scaling factor will cancel out.

Another way to deal with underflowing probabilities is to represent probabilities by their logarithms. This ensures that underflow cannot happen, and has the benefit that integers can be used to represent the logs, thereby changing floating point operations to fixed point ones. If we represent probability P with its log, $\log_b P$, we could get more precision by setting b closer to one. To multiply two numbers, we simply add their logarithms. Adding two numbers is more complicated. Let's assume that we want to add P_1 and P_2 and that $P_1 \geq P_2$:

$$\begin{aligned} &\log_b (P_1 + P_2) \\ &= \log_b [b^{\log_b P_1} + b^{\log_b P_2}] \\ &= \log_b [b^{\log_b P_1} (1 + b^{\log_b P_2 - \log_b P_1})] \\ &= \log_b P_1 + \log_b (1 + b^{\log_b P_2 - \log_b P_1}) \end{aligned} \tag{29}$$

Since integers are used to represent logarithms, if $\log_b (1 + b^{\log_b P_2 - \log_b P_1})$ is less than 0.5, the sum will simply be $\log_b P_1$. In other words, if P_2 is so many orders of magnitude smaller than P_1, adding the two numbers will just result in P_1. Moreover, if we could store all possible values of $\log_b P_2 - \log_b P_1$, the quantity $\log_b (1 + b^{\log_b P_2 - \log_b P_1})$ could be stored as a table, $T(n)$, where

$$T(n) = \begin{cases} \log_b (1 + b^n) & \text{if } T(n) \geq 0.5 \\ 0 & \text{otherwise} \end{cases} \tag{30}$$

The number of possible values depends on the magnitude of b. In our implementation, we chose $b = 1.0001$, which resulted in a table size of 99,041.

With the aid of this table,

$$\log_b (P_1 + P_2) = \begin{cases} \log_b P_1 + T(\log_b P_2 - \log_b P_1) & \text{if } P_1 > P_2 \\ \log_b P_2 + T(\log_b P_1 - \log_b P_2) & \text{otherwise} \end{cases} \tag{31}$$

This implements the addition of two probabilities as one integer add, one subtract, two compares, and one table lookup. Although using logarithm undoubtedly introduces errors, in practice, we have been able to obtain identical results when this representation is used recognition, and comparable results when it is used in training. [Brown 87] contains a more complete treatment and error analysis.

2.3.5. Multiple Independent Observations

The Baum-Welch algorithm has been described for one training datum, but is easily generalizable for multiple training sequences. This is particularly useful for speech recognition because often we would like to train a good model for some word (or phone) given a large number of observations of that word (or phone).

To train one set of HMM parameters from multiple samples, run the forward-backward algorithm on each independent sequence to calculate the expectations (or counts). These counts may be added together when all samples have been processed. Then all the parameters are re-estimated. This constitutes one iteration of Baum-Welch re-estimation. This procedure is maximum likelihood for the objective function:

$$\prod_{\forall d} \alpha_{S_F}(T_d) \tag{32}$$

where $\alpha_{S_F}(T_d)$ is the alpha-terminal of a training datum, d.

2.3.6. Smoothing

The amount of training data is usually insufficient with respect to the number of parameters in the HMM. As a result, the HMM parameters, particularly the output pdf's, may be inaccurate. While the frequently occurring output symbols will be well-trained, many other symbols will still be unobserved, and will have zero probability. In practice, some of the symbols with zero probability may occur in the future, and if they are left as zeroes, the input token will have a probability of zero. Therefore, some way of smoothing the output pdf is essential. There are two problems of interest with smoothing: (1) what to smooth the parameters with, and (2) how to smooth.

The simplest solution to the first problem is the *floor method*. All the zero probabilities are replaced with a very small value (typically around 10^{-5}). This was used by Levinson *et al.* [Levinson 83]. This technique efficiently solves the zero-probability problem and is sufficient for reasonably well-trained models. However, it cannot distinguish the unlikely output symbols from the impossible ones, which creates a problem when the models are not well trained and many codewords are not observed.

An improvement is the *distance method*. During training, multiple output symbols are provided and assigned probabilities based on their distances from the training vector and a Parzen-window or a K-nearest neighbor window. Then all of these labels are used for smoothing. Thus, if a codebook vector with a zero probability is similar to a codebook vector with a high probability, the probability of the first vector will be raised significantly. This technique has been used by several systems [Cravero 84, Schwartz 84].

A final approach is the *co-occurrence method* [Sugawara 85, Lee 88a], which uses information about which symbols are frequent replacements for each symbol. In other words, when a symbol is observed, how likely are the other symbols observed in similar contexts. This information could be derived from DTW of the same words [Sugawara 85], or directly from the output pdf's [Lee 88a]. From this co-occurrence frequency, a probabilistic mapping can be created and used to map individual output pdf's into smoothed pdf's.

These three approaches provide the means of finding alternative estimates, which can be smoothed with the trained parameters using a linear combination. For example, we could view the *floor method* as a linear interpolation between the trained output pdf and a uniform distribution. This interpolation is shown in Equation 33.

$$MP(i \mid d) = \lambda \cdot P(i \mid d) + (1-\lambda) \cdot SP(i \mid d) \tag{33}$$

where i is a codeword, d is the distribution being smoothed, P is the trained output parameter, SP is the smoothed parameters, and λ is an estimate of how good the trained parameters are in comparison to the smoothed parameters. λ can be estimated by trial and error when it is a single parameter [Levinson 83, Sugawara 85], but a better approach is to use λ's that depend on how well-trained each P is. When P is well-trained (large forward-backward count), a larger λ should be used. For this purpose, we could use *deleted interpolated estimation* [Jelinek 80], a technique that automatically determines λ depending on how well-trained P is. Basically, the two estimates are split into two parallel transitions, and the transition probabilities of these transitions can be trained by the forward-backward algorithm. These transition probabilities are the λ's. We will discuss deleted interpolation in more detail in Section 6.2.

2.4. Using HMMs for Speech Recognition

In this section, we will examine how HMMs can be used for speech recognition. We will discuss how to represent speech as output observations, how to construct models to represents units of speech, how to train HMMs from speech, and how to recognize speech with HMMs.

2.4.1. Representation

2.4.1.1. Continuous vs. Discrete Model

The description in the preceding section assume *discrete density HMMs*. With discrete HMMs, there are M output symbols, and the output probability density function, $P(Y_i=y_i \mid X_i=x_i, X_{i+1}=x_{i+1})$, is modeled explicitly. In order to apply these algorithms, each frame must be represented by a symbol from a finite alphabet. Vector quantization (VQ) [Linde 80, Makhoul 85] is an ideal method for this data compression. Basically, VQ tries to identify a set of prototype vectors from training data. Then, input speech is converted from a multi-dimensional real feature vector (of, say, FFT or LPC coefficients) to a symbol that represents the best-matching prototype vector. This technique will be discussed in more detail in Section 4.2.

The mathematical formulation presented earlier can be extended to cases where the observations are continuous multi-dimensional vectors. By assuming certain properties of these vectors' distributions, it is possible to estimate the output parameters from training data. The most frequently used continuous density is the multivariate Gaussian density [Paul 86]. With multivariate Gaussian density, an output pdf is described by a mean vector and a covariance matrix. To reduce computation, the covariance matrix is sometimes assumed to be diagonal (all the off-diagonal terms are zeroes). Other forms include the Gaussian mixture density [Rabiner 85], the Gaussian autoregressive mixture density [Juang 85b], the Richter mixture density [Richter 86], and the Laplacian mixture density [Ney 88].

The principal advantage of using a continuous HMM is the ability to directly model speech parameters, which are usually in the form of multi-dimensional real-valued feature vectors. One need not worry about vector quantization errors or selecting a general purpose distance metric in the vector quantizer. Also, the use of continuous parameters typically lead to a modest reduction in the number of parameters.

However, continuous HMMs require considerably longer training and recognition time. For example, using discrete HMMs, computing the output probability of an observation is merely a table-lookup. On the other hand, using continuous HMMs, many multiplies are required even with the simplest single-mixture, multivariate normal density with a diagonal covariance matrix. Yet, Rabiner *et al.* [Rabiner 85] showed that a single mixture diagonal covariance matrix does not adequately represent certain speech parameters. Brown [Brown 87] showed that full-covariance Gaussians or mixture Richters reduces the error rate of a diagonal covariance matrix by about 50%. But the use of multiple-mixture or full covariance further increases the complexity of both training and recognition with continuous HMMs.

Researchers have had mixed results comparing discrete and continuous HMMs. Bahl *et al.* [Bahl 81b] compared continuous density HMMs with discrete density HMMs on a 1000-word continuous speech task. Discrete HMMs led to a 10.5% error rate, while multivariate Gaussian continuous model with a diagonal covariance matrix resulted in a 21.9% error rate. More recently, Brown [Brown 87] showed that for E-set recognition, discrete density HMMs performed better than diagonal Gaussian continuous density HMMs. Other recent attempts at IBM [Jelinek 87] to replace discrete densities with continuous ones have not been successful.

On the other hand, Rabiner *et al.* [Rabiner 85] reported error rates between 0.7% and 2.4% for various continuous observation HMMs on isolated speaker-independent digit recognition, while discrete observation HMMs led to an error rate of 2.9%. Another study by Gupta *et al.* [Gupta 87] showed that for a 60,000-word isolated speaker-dependent task, the best version of discrete HMMs was 69% accurate, while full-covariance Gaussian models were 76% accurate.

Brown [Brown 87] provided excellent analyses to account for some of these discrepancies. He explained that maximum likelihood estimation (MLE) assumes that (1) the assumed distributions are correct, (2) the distributions are well-behaved, and (3) the sample size is large enough. When these assumptions fail, the behavior of MLE is no longer predictable. Therefore, when discrete HMMs are used, since their distributions are non-parametric and no assumptions are made, at least (1) and (2) would not be violated. His results also suggest that in order to use MLE and continuous parameters, accurate distributions (such as Gaussian mixture density or full covariance) are needed. However, these distributions require considerable computation.

Thus, while continuous HMMs have numerous advantages, their inefficiency, incorrect assumptions, and the lack of consensus in results led us to our decision to use discrete HMMs. Although discrete HMMs are less flexible, and cannot recover from vector quantization errors, they are extremely efficient. Furthermore, they can represent any distribution because no assumptions are made about the underlying distribution of the observed symbols. Finally, we do not have the resources to conduct extensive experiments with continuous density models. Thus, we believe that the advantages of discrete HMMs outweigh their disadvantages for our task, and will use discrete HMMs in this work. It should be noted, however, that continuous models are a very important area of research, and we expect that most of our ideas can be applied to continuous models.

2.4.1.2. HMM Representation of Speech Units

Hidden Markov models are a natural representation of speech. The output distribution models the parametric distribution of speech events, and the transition distribution models the duration of these events. HMMs can be used to represent any unit of speech. Since there are strong temporal constraints in speech, left-to-right models are usually used.

The most natural unit of speech is the word. Researchers at Bell Laboratories [Rabiner 85], Lincoln Labs [Lippmann 87], and IBM Japan [Nishimura 87] have used models similar to that in Figure 2-3. This type of model was first used by Bakis [Bakis 76]. In theory, a state could correspond to some phonetic event, and each event could be skipped. For example, the HMM in Figure 2-3 can be chosen to represent the word *did*, where each state represents a phonetic event. In practice, choosing the right number of events is tricky, and the optimal number does not usually agree with phonetician's intuition. Therefore, researchers usually choose more states than needed for the longest word, and use the same model for all words.

While words are just what we want to recognize, they are not a practical choice for large-vocabulary recognition because the amount of training and storage is enormous. Instead, some subword unit should be

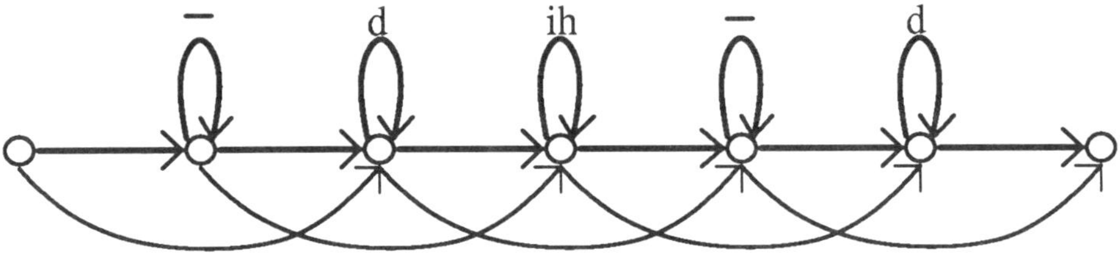

Figure 2-3: An HMM representing a word.

used. Researchers at CMU [Baker 75b, Lee 87], IBM [Bahl 80a], IBM-France [Derouault 87], BBN [Chow 87], and Philips [Noll 87] have used phone or phoneme models. An example of a phoneme model is shown in Figure 2-4. The three states with self-loops could represent the transition into the phoneme, the steady state portion, and the transition out of the phoneme.

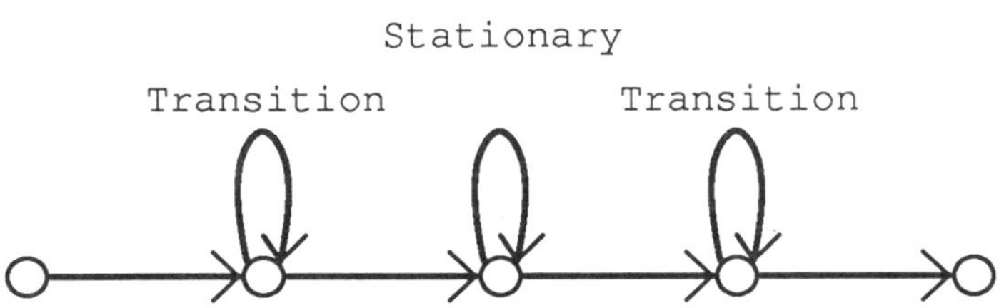

Figure 2-4: An HMM representing a phoneme.

It is, of course, possible to use HMMs to represent any other unit of speech, such as syllable, demisyllable, or diphone. In fact, hidden Markov models have an amazing ability to generalize even when poor units are selected. If the exact same HMM network is used for training and testing, any incorrect or suboptimal characteristics of the network are automatically absorbed into the *a* and *b* parameters. Bahl *et al.* [Bahl 83a] conducted an experiment where letter-by-letter spelling was used to model words. For example, the word *night* was represented as a concatenation of the models for *n, i, g, h,* and *t.* Naturally, the results from this baseform were considerably worse than that obtained using phonetic models. However, this representation proved to be "considerably better than the results obtained with the complete channel model using parameters estimated by people". This astounding result illustrates the power of automatic training and the importance of training and testing under the same conditions.

2.4.1.3. HMM Representation of Other Knowledge Sources

A central philosophy of HMM-based speech recognition is: *any knowledge source that can be represented as a hidden Markov model should be represented as one* [Baker 75a]. By representing all knowledge sources as HMMs, the recognition search merely consists of a search in an enormous HMM.

For a continuous word recognition task that uses no grammar and word HMMs, we could simply place all word models in parallel, and add an initial and final state. The initial state has a null transition to the initial state of each word model; the final state of each word model has a null transition to the final state. The final state has a null transition back to the initial state. The recognition search tries to determine the best path through this network. An example of a continuous digit recognition network is illustrated in Figure 2-5.

If we were to use phoneme models and a grammar, we could incorporate word and sentence knowledge into our recognizer in the following manner: Each word could be represented as a network of phonemes which encodes every way the word could be pronounced. The grammar could be represented as a network whose transitions are words, and the network would encode all legal sentences. We could then take the grammar network, instantiate each word with the network of phonemes, and then instantiate each instance of a phoneme with its hidden Markov model. Then we have a large HMM that encodes all the legal sentences. This is illustrated in Figure 2-6.

By placing all the knowledge in the data structures of the HMMs, it is possible to perform a global search that takes all the knowledge into account at every step. This *integrated search* is an important advantage that HMMs have over bottom-up or top-down systems.

2.4.2. Using HMM for Isolated Word Tasks

2.4.2.1. Training

HMM training for isolated words can be implemented directly using the forward-backward algorithm. First, collect many exemplars of each word in the vocabulary, then for each word, train an HMM from all the exemplars. It is not necessary to clip the speech from the beginning and ending silences

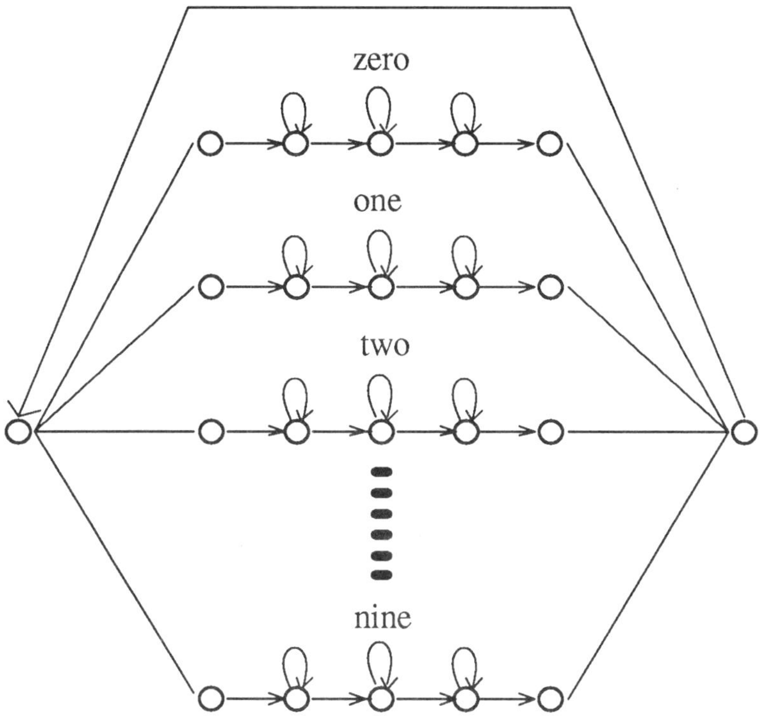

Figure 2-5: An HMM network for continuous digit recognition without grammar.

because they will be absorbed in the states of the word models. If subword units are used, these subword units should first be concatenated into a word model, possibly adding silence models at the beginning and end. Then the concatenated word HMM can be trained.

2.4.2.2. Recognition

Isolated word recognition is also easy to implement. We could use the forward pass to score the input word against each of the models. Assuming no language model, the model with the highest probability is chosen as the recognized word. We could also use the Viterbi algorithm for recognition, but the forward algorithm is preferred because the probability from Viterbi is only an estimate of the correct probability. If subword units are used, then they would be concatenated into words first.

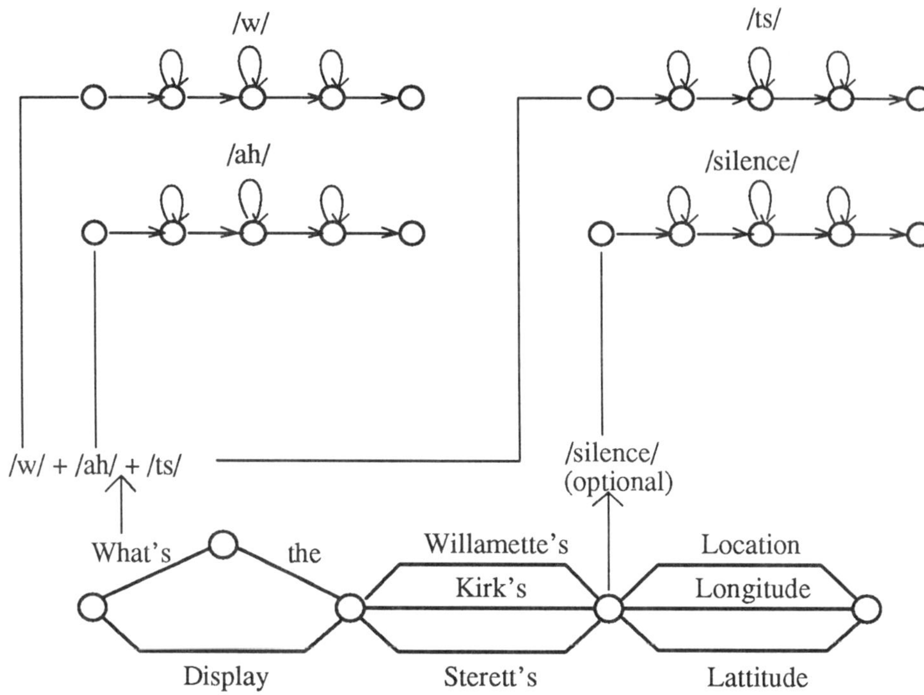

Figure 2-6: An HMM network for continuous word recognition with a finite state grammar.

2.4.3. Using HMM for Continuous Speech Tasks

2.4.3.1. Training

Techniques such as DTW, knowledge-engineering, traditional pattern recognition, and neural networks face serious problems in training their models for continuous speech, because word boundaries are not automatically detectable. Often, hand-marking is needed, which is tedious and may be suboptimal. HMMs, however, do not have this problem. In fact, training HMMs on continuous speech is not much different from training on isolated words.

Since the state sequence is hidden in HMMs, *it doesn't matter where the word boundaries are*. To train the parameters of HMMs, all we need is the word sequence in each sentence. Each word is instantiated with its model (which may be a concatenation of subword models). Next, the words in the sentence are concatenated with optional silence models between words. This large concatenated sentence HMM is then trained on the entire sentence. Since the entire sentence HMM is trained on the entire sentence, all word boundaries are considered at every frame. Only good state-to-speech alignments will have non-zero forward-backward counts.

This training algorithm for continuous speech has two advantages. First, no attempt is made to find word boundaries. The algorithm is allowed complete freedom to align the sentence model against the speech. Second, we make assumptions about how each word is pronounced. Some of these assumptions are surely inaccurate, but by training and testing on the same conditions, we are modeling our erroneous assumptions.

2.4.3.2. Recognition

With isolated word recognition, word boundaries are known, and only $N \times V$ searches have to be performed, where N is the number of words in the sentence and V is the size of the vocabulary. This is not the case for continuous speech recognition, where most word boundaries cannot be detected accurately. A word may begin and end anywhere, and all possible begin and end points have to be accounted for. This converts a linear search to a tree search, and a polynomial recognition algorithm to an exponential one.

However, while an optimal full search is infeasible for large-vocabulary continuous speech recognition, there are several good suboptimal searches that convert the problem back to a polynomial one.

Viterbi-based Searches

The Viterbi search [Viterbi 67] was already discussed as a solution to one of the three HMM problems in Section 2.2.2. To briefly reiterate, the Viterbi search is a time synchronous search algorithm that completely processes time t before going on to time $t+1$. For time t, each state is updated by the best score from states at time $t-1$. From this, the *most probable state sequence* can be recovered at the end of the search.

One way to extend the Viterbi search to continuous speech recognition is to first enumerate all the states of all the words in the grammar. For each frame, the above update is performed for within-word transitions, which are guaranteed to go from a lower-index state to a higher-index state. Then all between-word transitions specified by the grammar are tried. This second stage is needed, because between-word transitions are not guaranteed to be from lower-index states to higher-index ones, which violates the assumption in the Viterbi recursion. This is known as the *time-synchronous Viterbi search.*

A full Viterbi search is quite efficient for moderate tasks [Baker 75b]. However, for large tasks, it can be very time consuming. A very simple way to prune the search space is the *beam search* [Lowerre 76, Schwartz 85, Ney 87]. Instead of retaining all candidates at every time frame, a threshold T is used to consider only a group of likely candidates. The state with the highest log probability, p, is first found. Then each state with log probability $\leq p-T$ is discarded from further consideration. The use of the beam search alleviates the need to enumerate all the states, and can lead to substantial savings in computation with no loss of accuracy.

Another variation of the Viterbi search is *level building*, originally devised for dynamic-time-warp based recognition [Myers 81], and recently generalized to HMM-based recognition [Rabiner 88a, LeeCH 88a]. Level building is similar to Viterbi search in that it uses a time-state lattice, but it has an additional dimension—the number of words in a hypothesis. Level building first finds all one-word hypotheses from the beginning of the sentence. The best one-word hypothesis at each time frame is found. Then two-word hypotheses are found by extending every legal word from the best one-word hypotheses at every frame. This is repeated until some maximum level, L, is reached. Then the multi-word hypothesis with the best score is chosen as the recognized sentence.

This best level-building hypothesis is guaranteed to be the same as that obtained by the time-synchronous Viterbi search, as long as the solution from Viterbi is not longer than L words. The exhaustive Viterbi search extends one lattice of size $T \times S$, where T is the number of frames, and S is the number of states. On the other hand, level building extends L lattices of approximately the same size.

It may appear that level building can be useful when multiple hypotheses with different lengths are needed (possibly for postprocessing [Rabiner 88a]), or when a minimum, maximum, or known length constraints can be applied. However, if we use additional states in a time-synchronous Viterbi search, the same effects can be accomplished.

There are several advantages to using these Viterbi-based algorithms, which take the maximum at each state, rather than the forward algorithm, which sums over all incoming transitions at each state. First, Viterbi-based searches are more efficient. If we use a logarithmic representation of the HMM probabilities, Viterbi search is extremely efficient because multiplications of probabilities become additions. Second, it is possible to obtain the state sequence with the Viterbi algorithm, but not with the original forward algorithm. This could be important in many tasks, most notably continuous speech recognition. Although it is possible to extend the forward algorithm into stack decoding [Bahl 83a] for continuous speech recognition, the Viterbi algorithm will readily work for continuous speech recognition with almost no additional effort. Third, because of its time-synchronous nature, it is very easy to modify the time-synchronous Viterbi algorithm into a *beam search.* Such is not the case with most other searches.

There are also some disadvantages for using the Viterbi algorithm for recognition. Theoretically, the probability obtained from Viterbi is an approximation of the forward probability.[7] This is because the Viterbi search finds the *optimal state sequence* while speech recognizers should seek the *optimal word sequence.* Therefore, the Viterbi search is a suboptimal search. In view of this problem, Schwartz *et al.* [Schwartz 85] proposed a modification to the Viterbi search. Within words, transitions to the same state are added if they are within the same word. For between-word transitions, maximum is still used in order to find the best word sequence. This is an approximation of the forward algorithm, and is reported to produce better results. The problem with this algorithm is that it adds all paths within a word, even if they have different previous word sequences and begin/end times. Also, more time is needed to sum probabilities which have logarithmic representations.

Stack Decoding

Stack decoding [Bahl 83a] is a modification of the forward algorithm for continuous speech recognition. It is derived from the A* Search [Nilsson 80]. The A* search is not time-synchronous, but extends paths of different lengths. The search begins by adding all possible one-word hypotheses to the *OPEN* list. Then the *best* hypothesis is removed from the *OPEN* list, and

---

[7]In practice, the probabilities from forward and Viterbi are very close [Rabiner 86].

all paths from it are extended, evaluated, and placed back in the *OPEN* list. This search continues until a complete path that is guaranteed to be better than all paths in the *OPEN* list has been found. In selecting the *best* path, we need an evaluation function that estimates the score of the complete path as the sum of the known score of the partial path and the expected score of the remaining path. If the expected score of the remaining path is always an underestimate of the actual score, then the solution found will be optimal.

Several modifications to the A* search had to be made for speech recognition. First, an underestimating evaluation function is difficult to find for speech recognition. The ones that are guaranteed to underestimate will result in a very large *OPEN* list. So, a heuristic function that may over-estimate has to be used to prune more hypotheses. This invalidates the *admissibility* (optimality) of the algorithm.[8] Even with an over-estimating evaluation function, the *OPEN* list will still be much too large. So, two other types of pruning are used: (1) a fast-match that extends only a small fraction of paths from a partial path [Bahl 88a], and (2) the use of a *stack* (hence the name *stack decoding*) that saves only a fixed number of hypotheses in the *OPEN* list.

Viterbi search is a graph search, and paths cannot be summed because they may have different word histories. Stack decoding is a tree search, so each node in *OPEN* has a unique history, and the forward algorithm can be used within word hypotheses when the word is extended. With stack decoding, it is possible to use an objective function that searches for the optimal word string, rather than the optimal state sequence.

While the aforementioned properties of stack decoding are very attractive, many implementational problems arise for stack decoding. When partial paths of different lengths are allowed, we have to normalize their probabilities in order to compare them. This is very tricky since probabilities of different words or phones may not be comparable. In other words, while the acoustic probabilities being compared in Viterbi are always based on the same partial input, such is not the case in stack decoding. Another problem arises from the fact that hypotheses are extended one word at a time. It is necessary to sum over all word endings, which is time consuming. Thus,

---

[8]Although researchers have equated stack decoding to the A* Search, this is a misnomer. The A* Search *is* admissible, so stack decoding is really an example of *best first search*.

while stack decoding has many desirable properties, it is considerably more difficult to implement.

3 Task and Databases

3.1. The Resource Management Task and Database

We will be evaluating SPHINX on the *resource management* task [Price 88]. This task was designed for inquiry of naval resources, but can be generalized to database query. It was created to evaluate the recognizers of the recent DARPA projects, for example, CMU's speaker-independent ANGEL system [Adams 86], and BBN's speaker-dependent BYBLOS system [Chow 87].

3.1.1. The Vocabulary

At the lexical level, the 997-word resource management task is very difficult. There are many confusable pairs, such as *what* and *what's*, *what* and *was*, *the* and *a*, *four* and *fourth*, *are* and *were*, *any* and *many*, and many others. Most of the proper nouns can appear in singular, plural, and possessive forms, which creates more confusable pairs and ambiguous word boundaries. There are many function words (such as *a*, *and*, *of*, *the*, *to*), which are articulated very poorly and are hard to recognize or even locate. Moreover, many of these function words are optional according to the grammar. The entire vocabulary of this task, along with SPHINX's expected pronunciation of each word, are enumerated in Appendix II.1.

3.1.2. The Grammar

At the grammatic level, the resource management task is not a very difficult task because the sentences in this task are generated from a set of 900 sentence templates which resemble realistic questions in a database system. Some examples of these 900 templates are listed in Appendix II.2. The most obvious and correct way to model this language is to use a finite state language that generates the same set of sentences as these 900 templates [Baker 75b, Bahl 78a]. Such a grammar has been implemented by BBN, and resulted in 98.6% word accuracy for speaker-dependent recognition [Kubala 88]. However, this grammar has a perplexity[9] of about 9. Since the long-term goal of the DARPA effort is to investigate large-vocabulary high-perplexity tasks, this grammar is too simplistic. Moreover, with such high recognition rate, much more testing data will be needed to confirm the significance of an improvement. Therefore, DARPA suggested that looser grammars should be used with the resource management task.

To that end, we need a grammar that generates all sentences that could be generated by the 900 sentence templates, as well as some illegal sentences, so that the perplexity will be sufficiently high. BBN has proposed the *word pair grammar*, which is a simple grammar that specifies only the list of words that can legally follow any given word. This can be extracted from the 900 sentence templates. Each template is a network of "tags," or categories of words. Given these templates, we could easily determine what tags can follow any given tag. From this information and the list of words in each tag, we could find what words can follow any given word. Of the 994,009 word pairs, only 57,878 are legal: this grammar has a test-set perplexity of about 60.

To use this grammar for recognition, we could put the HMMs for each of the 997 words in parallel, and allow a null transition from the last state of word A to the first state of word B if (A,B) is a legal word pair. The transition probability of this null transition would be $\frac{1}{N}$, where N is the number of words that could follow word A.

Another grammar that could be constructed is the *bigram grammar.*

---

[9]Perplexity is roughly the number of choices per decision point, see Appendix I.1 for more details.

The bigram grammar also finds the list of words that can follow any given word, except instead of assigning equal probabilities to words, estimated probabilities are used. The probability that word W_2 follows word W_1 can be determined as follows:

$$P(W_2|W_1) \approx \sum_{\forall T_1} \sum_{\forall T_2} P(W_2|T_2) \cdot P(T_2|T_1) \cdot P(T_1|W_1) \tag{1}$$

where $P(T_2|T_1)$ is determined from the 900 templates by counting. But the values of $P(W_2|T_2)$ and $P(T_1|W_1)$ cannot be estimated from the 900 templates. Therefore, we assume that the probabilities of all words in a tag are equiprobable, and that the probabilities of all tags that a word belongs to are equiprobable.

The bigram grammar can be used in recognition exactly as the word pair grammar, except the probabilities of tag transitions are now estimated instead of assumed to be equal. The bigram grammar has a test set perplexity of about 20.

3.1.3. The TIRM Database

In addition to the aforementioned specifications, Texas Instruments supplied CMU with a large database of speech from this grammar. The TIRM database contains 80 "training" speakers, 40 "development test" speakers, and 40 "evaluation speakers." At the time of this writing, only the 80 training speakers and the 40 development test speakers are available. Among these speakers, 85 are male and 35 are female. Each speaker uttered 40 sentences from a list of 2700 sentences manually generated from the 900 templates (3 per template). The complete list of training and testing speakers can be found in Appendix II.3.

These sentences were recorded using a Sennheiser HMD-414-6, close-talking, noise-cancelling, headset-boom microphone in a sound-treated room. All speakers were untrained, and were instructed to read a list of sentences in a natural continuous fashion. The speech was sampled at 20 KHz at TI, downsampled to 16 KHz at the National Bureau of Standards and saved on magnetic tapes.

All 80 training speakers, as well as 25 of the development test speakers, were released to CMU for use as training material. The other 15 were released as interim test speakers for demonstrations in March and October of

1987. Ten sentences per speaker were designated as test sentences. Since the CMU's speaker-independent ANGEL system was tested on these 150 sentences, we divide these sentences as follows:

- 105 (80 training and 25 development test) speakers × 40 sentences each = 4200 training sentences.
- 15 speakers × 10 sentences each = 150 testing sentences.
- The same 15 speakers × 30 sentences each = 450 adaptation/tuning sentences.

In other words, we use all 80 training speakers plus 25 of the development test speakers as training data for SPHINX. We use the 10 designated sentences for each of the 15 remaining development test speakers as testing sentences under all conditions. The 30 other sentences for those 15 speakers can be used for speaker adaptation, or for tuning SPHINX. Appendix II.3 enumerates all 120 speakers.

3.2. The TIMIT Database

We have another database at our disposal, namely, the TIMIT (TI - MIT) [Lamel 86, Fisher 87] database. This database was constructed to train and evaluate speaker-independent phoneme recognizers. This database was recorded under exactly the same conditions as the TIRM database. It consists of 630 speakers, each saying 10 sentences, including:

- 2 "sa" sentences, which are the same across all speakers.
- 5 "sx" sentences, which were read from a list of phonetically balanced sentences selected by MIT.
- 3 "si" sentences, which were randomly selected by TI.

70% of the speakers are male. Most speakers are Caucasian adults.

These sentences were recorded, labeled, and made available to CMU in sets of 20 speakers. Prior to this study, CMU had received 14 sets, totalling 280 speakers, or 2800 sentences. For the use in this study, we chose not to use the "sa" sentences in training or recognition, because they introduce an unfair bias for certain phonemes in certain contexts. Some sentences or phonetic transcriptions were unreadable. Therefore, we actually have 2205 sentences at our disposal.

The MIT labels were modified slightly at CMU, leaving a set of 62 possible phonetic labels. These labels, along with examples, are enumerated in Table 3-1.

Phone	Example	Phone	Example	Phone	Example
/iy/	*beat*	/er/	*bird*	/z/	*zoo*
/ih/	*bit*	/axr/	*diner*	/zh/	*measure*
/eh/	*bet*	/el/	*bottle*	/v/	*very*
/ae/	*bat*	/em/	*yes'em*	/f/	*fief*
/ux/	*beauty*	/en/	*button*	/th/	*thief*
/ix/	*roses*	/eng/	*Washington*	/s/	*sis*
/ax/	*the*	/m/	*mom*	/sh/	*shoe*
/ah/	*butt*	/n/	*non*	/hh/	*hay*
/uw/	*boot*	/ng/	*sing*	/hv/	*Leheigh*
/uh/	*book*	/ch/	*church*	/pcl/	(p closure)
/ao/	*bought*	/jh/	*judge*	/tcl/	(t closure)
/aa/	*cot*	/dh/	*they*	/kcl/	(k closure)
/ey/	*bait*	/b/	*bob*	/qcl/	(q closure)
/ay/	*bite*	/d/	*dad*	/bcl/	(b closure)
/oy/	*boy*	/dx/	(bu*tt*er)	/dcl/	(d closure)
/aw/	*about*	/nx/	(flapped n)	/gcl/	(g closure)
/ow/	*boat*	/g/	*gag*	/epi/	(epin. clos.)
/l/	*led*	/p/	*pop*	/h#/	(beg. sil)
/r/	*red*	/t/	*tot*	/#h/	(end sil)
/y/	*yet*	/k/	*kick*	/pau/	(betw. sil)
/w/	*wet*	/q/	(glot. stop)		

Table 3-1: List of the phones used in the TIMIT database.

We used the TIMIT database in two ways. First, we built an HMM-based phoneme recognizer, and tuned various aspects of our representations based on recognition results. Since training and recognition are much faster, we are able to stabilize the basic building blocks of SPHINX more rapidly by tuning on phoneme recognition. This phoneme recognizer and results are described in [Lee 88b]. Second, we use the HMMs trained for phoneme recognition to initialize the HMMs used in word recognition. This proved to be an excellent initialization, and saved considerable time in actual training.

4 The Baseline SPHINX System

In order to establish a benchmark performance on the resource management task using standard HMM techniques, we begin with a baseline HMM system. This baseline system will use basic HMM techniques utilized by many other systems [Rabiner 83, Bahl 83a, Schwartz 84, Sugawara 85]. We will show that using these techniques alone, we can already attain reasonable accuracies.

4.1. Signal Processing

The speech is sampled at 16 KHz, and pre-emphasized with a filter whose transform function is $1 - 0.97z^{-1}$. This has the effect of spectral flattening.

The waveform is then blocked into frames. Each frame spans 20 msec, or 360 speech samples. Consecutive frames overlap by 10 msec, or 180 speech samples. Each frame is multiplied by a Hamming window with a width of 20 msec and applied every 10 msec.

From these smoothed speech samples we compute the LPC coefficients using the autocorrelation method [Markel 76]. LPC analysis was performed with order 14. Finally, a set of 12 LPC-derived cepstral coefficients are computed from the LPC coefficients.

This representation is very similar to that used by Shikano *et al.* [Shikano 86a] and Rabiner *et al.* [Rabiner 84]. Rabiner *et al.* [Rabiner 84] compared this representation against four other LPC-based representations, and found that the LPC cepstral coefficients yielded the highest recognition accuracy. Shikano [Shikano 86a, Shikano 86c] and Lee [Lee 85a] conducted similar experiments that led to the same conclusion.

4.2. Vector Quantization

Vector quantization (VQ) [Linde 80, Gray 84, Makhoul 85] is a data reduction technique that maps a real vector onto a discrete symbol. A vector quantizer is completely described by a *codebook*, which is a set of fixed *prototype vectors*, or reproduction vectors. Each prototype vector has the same dimensions as an input vector. To perform the mapping, the input vector is matched against each prototype vector in the codebook using some distortion measure. The input vector is then replaced by the index of the prototype vector with the smallest distortion.

In our application, an input vector corresponds to the set of 12 32-bit floating point LPC cepstrum coefficients, and is mapped to an 8-bit index which represents one of 256 prototype vectors. Thus, the speech data is reduced by a factor of 48. Yet, researchers have demonstrated that little or no accuracy is lost by vector quantization [Shikano 86a, Rabiner 84].

A description of the vector quantization process includes: (1) the distortion measure, and (2) the generation of the 256 prototype vectors.

4.2.1. The Distortion Measure

The distortion measure compares two frames of speech, and determines a distance that estimates the difference between them. The distortion measure used in the baseline SPHINX system is the standard cepstrum distance [Shikano 86d] as defined below:

$$CEP = \sum_{i=1}^{12} (C_i^r - C_i^t)^2, \tag{1}$$

where C_i^r and C_i^t represent the ith LPC cepstrum coefficient of the two frames being compared. This distance metric has been widely used, and provides a good starting point for our baseline system.

4.2.2. A Hierarchical VQ Algorithm

The goal of a vector quantization algorithm is to generate a number of prototype vectors from a large sample of training vectors. Our criterion in vector quantization is to select the prototype vectors to represent the distribution of the training vectors, and to minimize the total distortion of each training vector against the best matching prototype vector.

Our algorithm is a variant of the Linde-Buzo-Gray algorithm [Linde 80, Shikano 86a]. It is summarized in Figure 4-1. This algorithm iteratively splits the training data into 2, 4, 8, ... 256 partitions, with a *centroid* for each partition. The *centroid* is determined by iterative refinement: Each training vector is classified into the partition whose *centroid* best matches the vector. Then a new *centroid* is computed for each partition by averaging all of the training vectors in the partition. This iterative refinement procedure is said to have converged when the improvement of the average distortion compared to the distortion in the previous iteration falls below a pre-determined threshold, and the split stage follows. The split stage finds two points that are far apart in each partition using a heuristic method, and these two points are used as new centroids. The split stage doubles the number of partitions. Then, the labeling and centroid stages are iterated. Note that after a partition is split, a vector in it need not be in one of its offsprings. This algorithm terminates when the training vectors have been split into M partitions, and the M corresponding centroids have converged. These M centroids are stored as the prototype vectors of the VQ codebook.

In our application, 150,000 frames of non-overlapped 20-msec coefficients are used to generate a 256-vector codebook. 150,000 frames is larger than the typical training data size in order to characterize speech from many speakers. These frames were extracted from 4000 sentences (2000 from TIRM and 2000 from TIMIT) by taking about 40 frames from each sentence. In agreement with other studies [Schwartz 84, Shikano 86a], our preliminary experiments showed that 256 vectors produced the best recognition accuracy. This set of 256 vectors are the alphabet in the output pdf in our discrete HMMs.

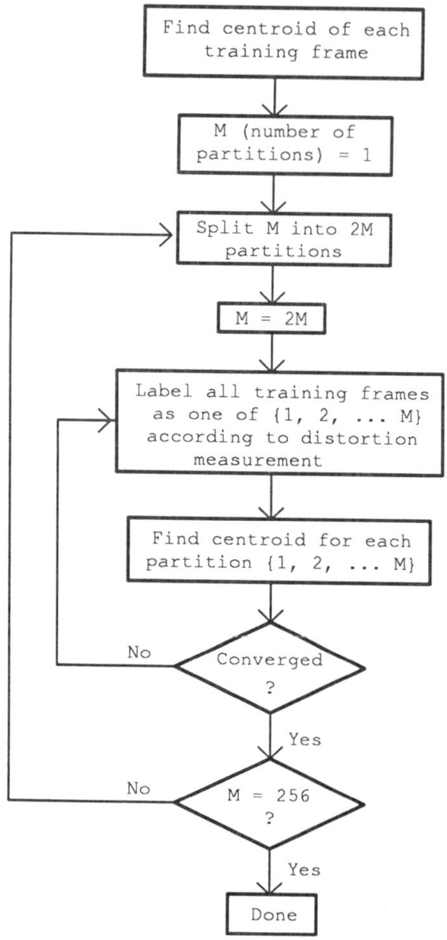

Figure 4-1: Flowchart of the vector quantization algorithm.

4.3. The Phone Model

We ran a set of preliminary phoneme recognition experiments that compared about ten different HMM topologies. We found that the best HMM topology is the one shown in Figure 4-2. This phone model is a slight modification of the model used in the phone-based version of IBM's TANGORA [Brown 87]. Our model has 7 states and 12 transitions. The transitions are tied into three groups (see Section 2.3.1). Transitions in the same group share the same output probabilities (represented by B, M, and E in the figure). This model assumes that there are at most three steady states for a phone, which are indicated by the self-loops. Furthermore, the lower states and transitions explicitly model short durations. Only one legal path exists for input of lengths 1, 2, 3, and 4. This has the advantage of better modeling of duration. With tied output transitions, these extra states and transitions do not demand more training data.

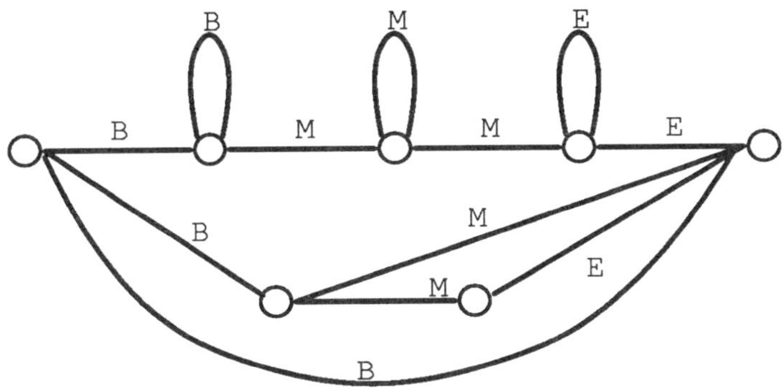

Figure 4-2: The phone HMM used in baseline SPHINX.

4.4. The Pronunciation Dictionary

For the baseline system, we used a slightly modified version of the baseforms from the ANGEL system [Rudnicky 87] as our pronunciation dictionary[10]. Each word is assumed to have one pronunciation, which is a sequence of phones. This sequence is very similar to what one might find in a dictionary. This dictionary uses a set of 45 phones, which are enumerated in Table 4-1. These phones are a subset of the phones shown in Table 3-1. The reason TIMIT phones were not directly used is because different labels were assigned to the same phoneme when their spectral properties were different. For example, /er/ represents stressed r and /axr/ represents unstressed r. Some of these phones are rare, and cannot be adequately trained. Also, with allophones for each phoneme, alternate pronunciations would have to be considered if TIMIT labels are used. Since we wanted to begin with a simple system with one pronunciation per word, we decided to map all allophones of a phoneme into a single phoneme. This mapping is also described in Table 4-1. Table 4-2 shows a section of our baseform dictionary. There are no phonological rules for inserting closures, or

---

[10]Note that this is not ANGEL's dictionary, but only the baseforms from which ANGEL dictionary was generated.

flapping /t/'s, /d/'s, and /n/'s.

Phone	Example	Allophones	Phone	Example	Allophones
/iy/	*beat*		/ng/	*sing*	/eng/
/ih/	*bit*		/ch/	*church*	
/eh/	*bet*		/jh/	*judge*	
/ae/	*bat*		/dh/	*they*	
/ix/	*roses*		/b/	*bob*	
/ax/	*the*		/d/	*dad*	
/ah/	*butt*		/g/	*gag*	
/uw/	*boot*	/ux/	/p/	*pop*	
/uh/	*book*		/t/	*tot*	
/ao/	*bought*		/k/	*kick*	
/aa/	*cot*		/z/	*zoo*	
/ey/	*bait*		/zh/	*measure*	
/ay/	*bite*		/v/	*very*	
/oy/	*boy*		/f/	*fief*	
/aw/	*bough*		/th/	*thief*	
/ow/	*boat*		/s/	*sis*	
/l/	*led*	/el/	/sh/	*shoe*	
/r/	*red*		/hh/	*hay*	/hv/
/y/	*yet*		/cl/	(unv. clos.)	/p,t,k,q-cl/
/w/	*wet*		/vcl/	(voi. clos.)	/b,d,g-cl/
/er/	*bird*	/axr/	/epi/	(epin. clos.)	
/m/	*mom*	/em/	/sil/	(silence)	/h#/,/#h/,/pau/
/n/	*non*	/en/, /nx/			

Table 4-1: List of the phones used in baseline SPHINX baseforms.

4.5. HMM Training

The baseforms in the baseline pronunciation dictionary use a total of 45 phones. We modeled each phone with an HMM. To initialize our phone model parameters, we used hand-segmented and -labeled segments from

Word	Pronunciation
ADDED	/ae d ix d/
ADDING	/ae d ix ng/
AFFECT	/ax f eh k t/
AFTER	/ae f t er/
AGAIN	/ax g eh n/
AJAX	/ey jh ae k s/
ALASKA	/ax l ae s k ax/
ALERT	/ax l er t/
ALERTS	/ax l er t s/

Table 4-2: Some baseform examples used in the baseline version of SPHINX.

2240 TIMIT sentences. We ran one iteration of forward-backward on these hand-labeled phone segments, and produced a model for each phone. This set of 45 phone models was used to initialize the parameters in the actual training.

In the actual training, we first replaced all output probabilities less than 0.001 with 0.001 in order to mitigate the effect of erroneous initial phone model parameters, and to prevent mimicking the hand segments in a different task. After this initialization, we ran the forward-backward algorithm on the resource management training sentences. For each of the 4200 sentences, we created a sentence model from word models, which were in turn concatenated from phone models. In addition to connecting word models together, we accounted for silences by inserting a mandatory silence at the beginning and at the end of the the sentence. Between-word silences were also allowed, but could be skipped. The skip probability was fixed at 0.5. This sentence model represents the *expected pronunciation* of this sentence. It was trained against the actual input speech using the forward-backward algorithm.

Two iterations of forward-backward training were run. Most other HMM systems run more iterations; however, we found that with our appropriate initialization, the joint probability of generating all training sentences increased very slightly after two iterations. Moreover, empirical results showed that recognition did not improve after two iterations. After four iterations, the recognition rate actually deteriorated slightly. This might be because the models have been made "too suitable" for the training data, and can no longer recognize some cases where the test data deviate from the training data.

The trained transition probabilities were used directly in recognition. The output probabilities, however, were smoothed using *deleted interpolation*, as described in Section 2.3.6. During the final iteration, the training data was divided into two blocks, and separate counts were maintained. After the final iteration, we combined each output pdf with a uniform distribution (or $\frac{1}{256}$ for each output probability) using weights that were dependent on the count of that output pdf. This has the effect of smoothing poorly-trained pdf's more than well-trained ones. The λs used to smooth the trained pdf's and a uniform distribution are shown in Table 4-3.

Count Range	$\lambda_{HMM\ params.}$	$\lambda_{Uniform\ dist.}$
$300 > Count \geq 0$	0.819	0.180
$700 > Count \geq 300$	0.925	0.075
$1250 > Count \geq 700$	0.957	0.043
$2000 > Count \geq 1250$	0.976	0.024
$3500 > Count \geq 2000$	0.986	0.014
$7000 > Count \geq 3500$	0.996	0.004
$\infty > Count \geq 7000$	0.998	0.002

Table 4-3: λs trained by deleted interpolation to smooth trained HMM parameters with a uniform distribution.

We found that smoothing only resulted in small improvements, because with 4200 sentences and only 45 phone models, our HMM parameters were very well trained. In subsequent experiments with more models, smoothing became more important. Without smoothing, recognition is typically 30-50% faster because more paths are pruned, and the recognition rate is the same for most sentences; however, occasionally a sentence would be very poorly recognized, or all paths may be pruned due to grammar. Therefore, smoothing is still very important, although it did not make much difference in the baseline system.

4.6. HMM Recognition

The SPHINX recognition search is a Viterbi beam search (see Section 2.4.3.2). The search processes input speech time-synchronously, updating all accessible states for a time frame $t-1$ completely before moving on to frame t. The update for time t consists of two stages.

First, for each within-word transition between states s_{from} and s_{to}, if $P(s_{from}, t-1) \cdot P(transition) \cdot P(output)$ is greater than $P(s_{to}, t)$, then $P(s_{to}, t)$ is updated. Second, for the final state of every word, try all legal word successors. Unlike the first stage, no output is emitted, so the update involves only $P(s, t-1) \cdot P(transition)$, where $P(transition)$ is the probability of the word we are about to enter according to the language model. If we don't use any language model, $P(transition) \approx \frac{1}{Vocabulary\ size}$. If we use a word pair or bigram grammar, $P(transition) \approx P(word_2 \mid word_1)$.

It appears that this probability combination algorithm should work; however, in practice, due to the fallacy of the Markov and independence assumptions, the probability of acoustics is underestimated. Combining a "normal" language model probability with an underestimated acoustic model probability would give the language model too little weight. In order to make them comparable, a *language model match factor* must be introduced. Then, by raising the language model probability to that power, the two scores are more balanced. Another way to look at this problem is to view the language model probability as a penalty for exiting words. If this penalty is large, the recognizer would prefer longer words, and if this penalty is small, the recognizer would prefer shorter words. The same mechanism was used in the IBM recognizer [Bahl 80b].

In all the results that we present in this work, we have used a fixed *language model match factor* for each grammar, which was originally obtained from tuning data. We have tried to vary this factor for different system configurations, but always found that the original choices to be optimal, or nearly optimal.

In the Viterbi beam search, a hypothesis is pruned if its log probability is less than that of the best hypothesis by more than a preset threshold. We found it is possible to prune 80–90% of the hypotheses, without any loss in accuracy.

When the search is completed, a backtrace is performed to recover the best path. This is done by keeping a history pointer at each node. The history pointer points to the previous word, and remembers the beginning time of the current word.

4.7. Results and Discussion

The results with the baseline SPHINX system, using 15 new speakers with 10 sentences each for evaluation, are shown in Table 4-4. Percent correct is the percent of words recognized correctly, and word accuracy is percent correct minus percent of insertions. These figures were computed using a dynamic programming string match algorithm supplied by the National Bureau of Standards (see Appendix I.2 for more details). The results for recognition without grammar count homonym confusions (such as *ship's* and *ships*, or *two* and *too*) as correct. Such confusions are considered errors when a grammar is used. This method of error counting is used throughout this study.

Grammar	Perplexity	Percent Correct	Word Accuracy
None	997	31.1%	25.8%
Word-Pair	60	61.8%	58.1%
Bigram	20	76.1%	74.8%

Table 4-4: Baseline SPHINX results, evaluated on 150 sentences from 15 speakers.

The word accuracy improved substantially with the use of grammar. With a word-pair grammar of perplexity 60, the recognition rate more than doubled, and with a bigram grammar of perplexity 20, the word accuracy almost tripled.

These results can be directly compared to those of the ANGEL System, a large-vocabulary speaker-independent continuous speech recognition system based on knowledge engineering developed at CMU. In fact, the same 150 sentences were used to evaluate ANGEL. The results for 60 of the 150 sentences attained by ANGEL are shown in Table 4-5. The remaining 90 sentences were evaluated using an earlier and inferior version of ANGEL, so only results on the 60 are reported here. ANGEL used a trigram grammar, which estimated the probability of a word given the two previous words. This trigram grammar had a higher perplexity than our bigram grammar because a very large floor value was imposed. From the results in Table 4-5, we see that while the results with the baseline SPHINX were by no means impressive, they have already surpassed that of the ANGEL System.

Grammar	Perplexity	Percent Correct	Word Accuracy
Trigram	34	45.5%	41.0%

Table 4-5: Results of the speaker-independent ANGEL System on the same task using a subset (6 speakers) of the SPHINX data.

BBN's *speaker-dependent* BYBLOS System, also based on hidden Markov modeling, has been evaluated on the same task. The results of BYBLOS are shown in Table 4-6. These results were the average of 200 sentences from 8 speakers. Each speaker was trained on about 600 sentences of his/her speech, and tested on another 25. All 8 speakers were different from those used to evaluate ANGEL and SPHINX. We expect that with so many test speakers for both systems, the variations due to speakers should be minimal, although we expect to be testing on the same speakers in the future. Compared to BYBLOS, our baseline results are much worse. The only major difference between BYBLOS and our baseline system is the use of context-dependent modeling; however, if we extrapolate the results in [Chow 86], it is clear that even if BYBLOS were to use context-independent phone models, it would still have a much lower error rate than the baseline SPHINX. This comparison between the speaker-*dependent* BYBLOS System and the speaker-*independent* baseline SPHINX System confirms the three to five time error rate increase that many researchers use as a rule of thumb. It is this large difference that we will try to overcome in the succeeding chapters.

Grammar	Perplexity	Percent Correct	Word Accuracy
None	997	70.1%	67.6%
Word-Pair	60	94.8%	92.5%

Table 4-6: Results of the speaker-dependent BYBLOS System on the same task (200 sentences from 8 speakers.)

4.8. Summary

In this chapter, we have described a baseline implementation of SPHINX, which was based on standard HMM technology. We used vector quantization and discrete HMMs for expedience. The baseline SPHINX system was based on phone modeling. Each phone was represented by an HMM that used three distributions to model the beginning, middle, and end of a phone, and seven states to explicitly model short durations. Recognition was carried out by a Viterbi beam search.

The results of this system are mediocre. We report 26%, 58%, and 75% accuracy for grammars with perplexity 997, 60, and 20. Since the perplexity-20 grammar is already a tight grammar, we conclude that the performance of our baseline system is inadequate for any realistic large-vocabulary applications.

5 Adding Knowledge

Although human knowledge of speech is far from perfect, most researchers have much more knowledge than their speech recognizers. However, automatic speech recognizers perform reasonably despite their ignorance because of their superior ability to quantify the little knowledge they have, and to apply all the constraints they can.

It seems natural to try to put more human knowledge into speech recognizers. Many researchers advocate the *knowledge-engineering* approach to speech recognition [Haton 84, Zue 85, Cole 86b, Thompson 87]. The knowledge engineering approach takes the opposite view of the learning approach. While hidden Markov learning places learning entirely in the *automatic training algorithm*, the knowledge engineering approach attempts to *explicitly program heuristic knowledge* about acoustic/phonetic events into the recognizer. Whereas an HMM-based search is data-driven, a knowledge engineering search is typically heuristically guided.

There are several problems with a purely knowledge engineering acoustic-phonetic approach to speech recognition. An acoustic-phonetic approach to word recognition usually detaches the acoustic-phonetic component from the system. If this detachment is bottom-up, errors made in earlier components cannot be recovered, and knowledge of different components cannot be used in unison. If the various components communicate through a blackboard [Lesser 75], then extremely complex communication mechanisms must be used. Up to now, such mechanisms have underperformed integrated search. There are also other problems, such as excessive dependence on human guidance, and the lack of a trainable and tractable model.

Alternatively, we could use speech knowledge to enhance a speech recognizer within the framework of stochastic modeling. In this chapter, we discuss how we enhanced SPHINX with fixed-width speech parameters, which use very crude human knowledge, and with variable-width speech parameters, which utilize more human knowledge, but are difficult to integrate. We also discuss various ways of integrating these parameters. Finally, we will examine how the use of lexical and phonetic knowledge can improve the performance of SPHINX.

5.1. Fixed-Width Speech Parameters

The easiest way to add more knowledge to HMMs is to introduce more fixed-width parameters, or parameters that can be computed for every fixed-size frame. All we have to do is to devise a way of incorporating these parameters into the output pdf of the HMMs. In this section, we will consider several types of frame-based parameters, and discuss several ways of integrating them.

5.1.1. Bilinear Transform on the Cepstrum Coefficients

Although the LPC cepstrum coefficients have been shown to be an excellent representation for speech recognition [Rabiner 84, Shikano 86c], they are distributed along a linear frequency axis. This is undesirable because the ability of the human ear to discriminate between frequencies is approximated by a logarithmic function of the frequency, or a *bark scale* [Zwicker 61]. Furthermore, Davis and Mermelstein [Davis 80] demonstrated mel-scaled coefficients yield superior recognition accuracy compared to linearly scaled ones. Therefore, there is strong motivation for transforming the LPC cepstrum coefficients into a mel-frequency scale.

Shikano [Shikano 86c] applied a *bilinear transform* [Oppenheim 72] to speaker-dependent phonetic unit recognition. He reported significant improvements with the addition of bilinear transform. Bilinear transform is a technique that transforms a linear frequency axis into a warped one using an all-pass filter shown in Equation 1 and 2.

$$z_{new}^{-1} = \frac{(z^{-1} - a)}{(1 - a z^{-1})}, \quad (-1 < a < 1) \tag{1}$$

$$\omega_{new} = \omega + 2 \tan^{-1}\left(\frac{a \sin \omega}{1 - a \cos \omega}\right) \tag{2}$$

where ω is the sampling frequency expressed by the normalized angular frequency, ω_{new} is the converted frequency, and a is the frequency warping parameter. Positive a converts the frequency axis into a low-frequency weighted one by lengthening the low-frequency axis and shortening the pre-frequency axis. When a takes on values between 0.4 and 0.8, the frequency warping by a bilinear transform is comparable to that of the mel or Bark scales. In this work, a value of 0.6 was used for α.

5.1.2. Differenced Cepstrum Coefficients

Temporal changes in the spectra are believed to play an important role in human perception [Ruske 82]. Rising or falling formant slopes are an important cue in human spectrogram reading. Differential parameters are particularly useful in speaker-independent recognition because while absolute formant locations may be shifted for different speakers, formant slopes are relatively invariant across speakers. Thus, it would be desirable to incorporate "slope" measurements into recognizers. Moreover, since HMMs assume each frame is independent of the past, it would be desirable to broaden the scope of a frame.

Most speech recognition systems use only instantaneous coefficients. Only recently have researchers begun to use coefficients that measure dynamic changes in the spectra. Furui [Furui 86] used linear *regression coefficients* to measure the change in spectra. The *regression coefficient*, essentially a slope measurement, is defined as:

$$R_m(t) = \frac{\sum_{n=-\delta}^{\delta} n\, C_m(t+n)}{\sum_{n=-\delta}^{\delta} n^2} \tag{3}$$

where $C_m(t)$ is the mth coefficient of the tth frame of the utterance, and R is the corresponding regression coefficient. The slope is measured from $-\delta$ to δ. Rabiner *et al.* [Rabiner 88a] used the same set of coefficients, which they called *delta cepstrum.*

Shikano [Shikano 86c], on the other hand, suggested that the regression coefficients have too strong a slope-smoothing effect and are more expensive computationally. He proposed the use of *differenced LPC cepstrum coefficients*, which is simply computed by:

$$D_m(t) = C_m(t+\delta) - C_m(t-\delta) \tag{4}$$

A similar measurement was used by Paul *et al.* [Paul 86].

We also conducted a preliminary experiment comparing the regression and the differenced coefficients for speaker-independent phoneme recognition. Our result showed that differenced coefficients are slightly better. In view of their computational simplicity, we decided to use the differenced coefficients. In our current implementation, a differenced coefficient is computed for every frame, with $\delta = 20$ msec, or 2 frames. This 40 msec difference produced the best result in a preliminary experiment.

5.1.3. Power and Differenced Power

Although LPC-based parameters performed well in speech recognition, they do not contain sufficient information about power. For example, coefficients in silence or noise regions are not very meaningful. Consequently, an LPC-based system may be capable of distinguishing vowels, but may have trouble separating speech from silence, a much easier task. Therefore, it is desirable to incorporate power into our recognizer. Rabiner, *et al.* [Rabiner 84] obtained significant improvement by adding power into the distance metric in vector quantization. Shikano [Shikano 85] reported similar results. Finally, in a detailed study of using prosody in speech recognition, Waibel [Waibel 86] found power to be the most important prosodic cue.

Power can be simply computed from the waveform as:

$$P = \log\left(\sum_{i=1}^{M} x_i^2\right) \tag{5}$$

where P is the power for frame n, which has M discrete time samples in it, namely $x_1, \cdots, x_M$. In our case, x_i's have been Hamming windowed.

Power itself is not a reliable source of information, because the absolute power of two speakers may be quite different. In order to normalize for speaker loudness variations, we could use an estimated maximum for the speaker to normalize his speech. In this study, we simply subtracted the maximum power value in the sentence from each power value in the sentence. In a real-time system, however, we cannot afford the delay required to estimate the speaker's maximum. Instead, some type of automatic gain control algorithm with minimal look-ahead could be used to predict the maximum power in a sentence.

Another important source of information is *differenced power*, which is computed the same way as *differenced LPC cepstrum coefficients*. Differenced power provides information about relative changes in amplitude or loudness. Our preliminary experiments indicated that differenced power is actually more useful than power.

5.1.4. Integrating Frame-Based Parameters

In the preceding sections, we introduced several frame-based parameters that should improve SPHINX's performance. Now, we will discuss three methods of integrating these parameters into SPHINX.

5.1.4.1. Stack and Reduce

The earliest effort that used differential information was described by the IBM speech group [Nadas 81]. They incorporated differential information not by differenced or regression coefficients, but by concatenating adjacent frames together. This doubled the number of parameters and introduced estimation problems. Their solution was to reduce the parameter size using principal component analysis [Duda 73].

Earlier work at IBM [Nadas 81, Bahl 83b, Das 83] described many variations of stack and reduce. More recently, Brown [Brown 87] applied discriminant analysis instead of principal components to reduce the dimensionality for E-set recognition. However, the IBM Speech Recognition Group has abandoned the stack and reduce approach for large-vocabulary recognition [Picheny 88], because it did not lead to significant improvements.

Differenced coefficients assume that spectral slope is the only desired feature, and nothing else is extracted. The stack and reduce approach does not make such an assumption, and is capable of finding all pairwise relationships from principal component analysis. Another advantage is that after principal components, the new features are uncorrelated and have equal variance, and a simple Euclidean distance can be used.

There are, however, several shortcomings of the stack and reduce approach. First, this approach requires more computation. Second, principal components require the estimation of a covariance matrix, and the off-diagonal terms are typically noisy, which may cause problems. Finally, principal component analysis is variance-maximizing, and may not yield optimal features for recognition or discrimination.

5.1.4.2. Composite Distance Metric

Another method for combining these parameters is to use a *composite distance metric*, or a weighted Euclidean distance. This approach was adopted by Furui [Furui 86] and Shikano [Shikano 86b]. To combine LPC cepstrum coefficients, differenced LPC cepstrum coefficients, power, and differenced power, we could use the following distance metric:

$$PDCEP = \sum_{i=1}^{12} (C_i^r - C_i^t)^2 + W_d \sum_{i=1}^{12} (D_i^r - D_i^t)^2$$
$$+ W_p (C_0^r - C_0^t)^2 + W_{p'} (D_0^r - D_0^t)^2 \qquad (6)$$

where C_i represents an LPC cepstrum coefficient, D_i is a differenced LPC cepstrum coefficient, C_0 is the power term, and D_0 is the differenced power. W_d, W_p, and $W_{p'}$ are empirically determined weighting factors that account for the relative importance and magnitude of differenced coefficients. In Shikano's study [Shikano 86c], best recognition accuracy was achieved with the following weight vector: {0.5, 0.0, 0.03}.[11]

We have performed similar phonetic recognition experiments [Lee 88a], and found the optimal weight vector to be: {0.8, 0.01, 0.05}. This discrepancy is due to the fact that Shikano's task was speaker dependent, where instantaneous parameters (C) are more reliable, and our task is speaker independent, where differential parameters (D) have increased importance.

Although the composite distance metric is efficient and known to

---

[11]Power was not used in Shikano's study.

perform well [Lee 85b, Lee 85a, Shikano 86c, Shikano 86a, Tohkura 86], it results in very large vector quantization distortion, or VQ error, due to the large number of dimensions in which VQ is performed. This is a very serious problem. Moreover, using *ad-hoc* linear weights is mathematically unappealing.

5.1.4.3. Multiple Codebooks

A final alternative, proposed by Gupta *et al.* [Gupta 87], is to quantize each set of parameters into a separate codebook. For each frame of speech, not one but several VQ codewords would be used to replace the input vector. Since each input frame is no longer a single symbol, but rather a vector of symbols, the discrete HMM algorithms must be modified to produce multiple symbols at each time frame. One way to combine these multiple output observations is to assume that they are independent. The output probability of emitting multiple symbols can then be computed as the product of the probability of producing each symbol. In other words, the α computation could be modified as in Equation 7:

$$\alpha_i(t) = \sum_j \{ \alpha_j(t-1) a_{ji} \prod_c b_{ji}(y_t^c) \} \qquad (7)$$

where c is one of the codebooks used. An example of a three-codebook, three-distribution HMM for the phone /ae/ is shown in Figure 5-1. The first and second codebooks correspond to the cepstrum coefficients and the differenced cepstrum coefficients. The third codebook is a combination of power and differenced power. Since both features are one-dimensional, we chose to combine them in one codebook using a composite distance metric. The codewords for power and cepstrum are sorted by power, and the codewords for differenced cepstrum are sorted by differenced power.[12] We can see that for the cepstrum and power codebooks, the middle distribution is the sharpest and loudest. For the differenced cepstrum codebook, the first coefficient is rising in the beginning, falling in the end. Since /ae/ is one of the loudest phones, this model appears plausible.

The multiple codebook approach has a distinct advantage over the

[12] Although power was not used in the cepstrum codebook, and differenced power was not used in the differenced cepstrum codebook, when we generated the cepstrum and the differenced cepstrum codebooks, we carried power and differenced power along solely for the purpose of sorting the codewords.

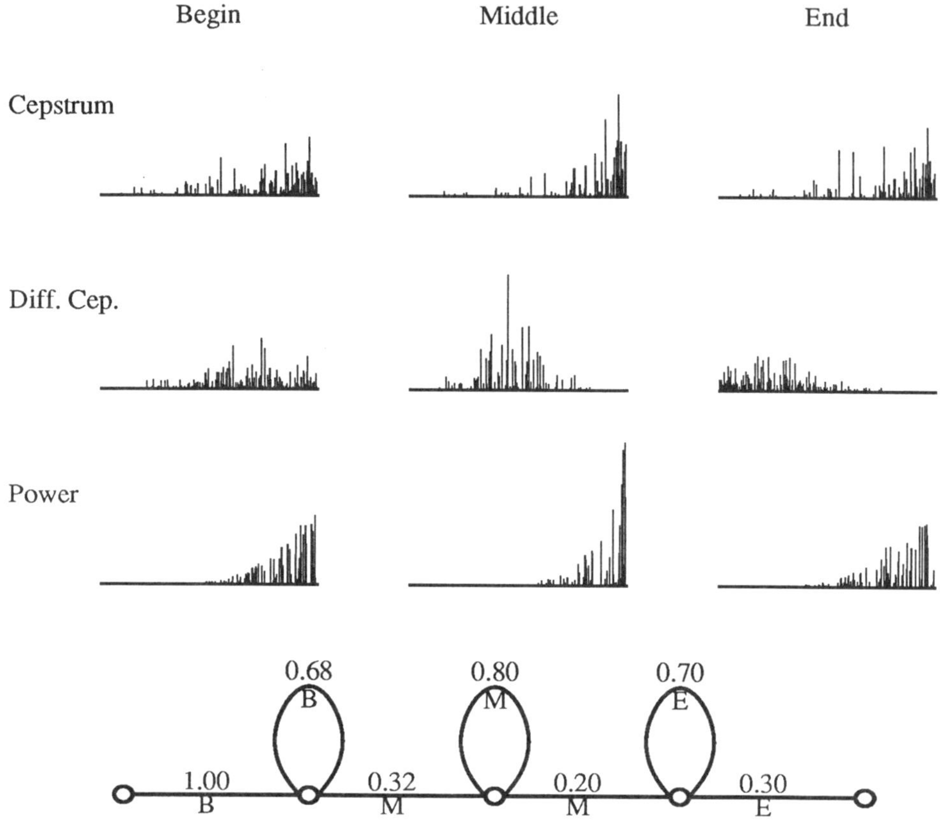

Figure 5-1: A phonemic hidden Markov model for phoneme /ae/. The upper portion displays the 9 output pdf's, where the x-axis is the codeword index, sorted by power and differenced power, and the y-axis is probability. The lower portion shows the HMM topology with transition probabilities. Transitions with the same label are tied to the same output pdf.

composite distance metric approach—namely, the reduction of quantization error. With the composite distance approach, the distortion is often quite large, which means the observed vectors match their corresponding prototype vectors poorly. When the dimensionality is increased as in the composite distance approach, the quantization error increases significantly. Table 5-1 illustrates the average distortion for several codebooks: (1) 12 stationary LPC cepstrum coefficients (2) 12 differenced LPC cepstrum coefficients, (3) differenced power and power, (4) the total distortion of these three codebooks weighted as in Equation 6, and (5) combination of all 26 coefficients into one codebook using the same weights. From this table, we see that the splitting stage in VQ reduces VQ error much faster when fewer features are used. Since we used the same weights in combining distortions of the separate codebooks and in the distance metric of the single codebook, these distortions are directly comparable. We see that the distortion of three size-16 codebooks is equivalent to that of a single size-256 composite codebook. This demonstrates that quantization error is significantly reduced by partitioning the parameters into separate codebooks.

Codebook Size	cep. dist.	dcep. dist.	power dist.	total dist.	1-codeb. dist.
2	0.84	0.93	3.99	1.86	2.42
4	0.57	0.56	1.39	1.12	1.94
8	0.42	0.43	0.60	0.81	1.45
16	0.32	0.34	0.29	0.61	1.19
32	0.25	0.28	0.15	0.48	1.00
64	0.20	0.23	0.07	0.39	0.83
128	0.16	0.19	0.03	0.31	0.72
256	0.13	0.15	0.01	0.25	0.61

Table 5-1: Quantization error of smaller codebooks, the weighted quantization error of three codebooks, and the quantization error of the combined codebook. This illustrates that quantization error is significantly reduced when parameters are partitioned into multiple codebooks.

Another advantage is the large increase in the dynamic range and precision of the multiple codebooks. With three codebooks, there are 256^3 possible parameter combinations with just 256×3 parameters. With such increase in precision comes the ability to make finer distinctions.

We further hypothesize that multiple codebooks are particularly suitable for well-trained speaker-independent models. Speaker-independent models have flatter distributions, and are more robust but less accurate than speaker-dependent models. With multiple codebooks, we multiply three relatively flat distributions and get a much sharper distribution. Moreover, since the original distributions are robust, minor deviations in quantization or speech realization still would not be disastrous, whereas in the speaker-dependent case, where more output probabilities are close to zero, the models become very sensitive to slight deviations from the norm. Therefore, multiple codebooks increase the ability of speaker-independent HMMs to make fine distinctions, yet retain sufficient robustness to deal with deviations.

One problem with the multiple codebook approach is the need for substantially more storage. In our case, we approximately tripled the number of parameters in our models.[13]

Using multiple codebooks will, of course, increase the time required for recognition. Although the modification occurs in the innermost loop, the total increase in computation is only about 10% because (1) most of the time is spent in traversing data structures, and (2) multiplications are implemented as adds.

5.2. Variable-Width Speech Parameters

HMM procedures process input frames time-synchronously. Typically, these frames have fixed width. One could add variable-width parameters by converting them to fixed-width ones or by pretending they are fixed-width. However, results with these approaches have not been promising [Bahl 78b, Nag 86]. Therefore, we chose to separately train variable-width features, and integrate them into the HMM search in an *ad-hoc* combination.

5.2.1. Duration

HMMs model duration of events with transition probabilities, which leads to an exponential distribution for the duration of state residence, for states with self-loops:

$$P_i(d) = (1 - a_{ii})\, a_{ii}^t \qquad (8)$$

where $P_i(d)$ is the probability of taking the self-loop at state i for exactly t times. It has been argued that this is an inadequate distribution for speech

[13]Note that increased parameters do not create estimation problems because the same training data is repeatedly used for the three separate codebooks.

events. One alternative is the use of *semi-Markov models* [Russell 85], which models not only the output and transition probabilities, but also a set of state duration probabilities: $P_i(d)$, for $D \geq d \geq 0$. By modifying the forward-backward algorithm, $P_i(d)$ can be estimated along with the other HMM parameters. Thus, semi-Markov modeling is mathematically appealing. However, this requires a D^2-fold increase in computation, and a D-fold increase in space, because for every non-self-transition, there are D transitions, and each of these transitions requires the multiplication of up to D output probabilities. This increase in computation is not warranted, since duration modeling will provide a modest improvement at best.

A more efficient, but suboptimal, approach is to estimate the HMM parameters and $P_i(d)$ separately, and combine them during recognition. While this deprives the optimality of the Viterbi algorithm, it has led to results equivalent to semi-Markov models while reducing computation significantly [Rabiner 85]. Rabiner *et al.* [Rabiner 85, Rabiner 88a] used $P_i(d)$ as a postprocessor to re-rank complete hypotheses from a level-building algorithm. In another study, Lowerre and Reddy [Lowerre 80] used a maximum and a minimum to constrain the duration of a phoneme.

We chose to apply our duration module as a part of the search, rather than as a postprocessor. For example, if we were to model state residency, we would first estimate $P_i(d)$ for $D \geq d \geq 0$. During recognition, every time state i is exited, $P_i(d)$ is multiplied with the transition and output probabilities. To normalize for different dynamic ranges, $P_i(d)$ is raised to a power, which is empirically determined. We call this *segment-level integration*, which is shown in Figure 5-2.

We have considered duration modeling for three types of events: (1) state duration, (2) phone duration, and (3) word duration. Modeling state and phone duration is time consuming, and particularly so when all HMM information is compiled into a large network, in which information about the current phone or state is difficult or costly to recover. Moreover, recall that our phone HMM already explicitly models duration of short events with the five lower transitions. Therefore, only word duration modeling was implemented.

We created a duration model for words that estimated [Hon 88]:

$$P_w(d), \quad \mu_w + 25 \geq d \geq \mu_w - 25 \qquad (9)$$

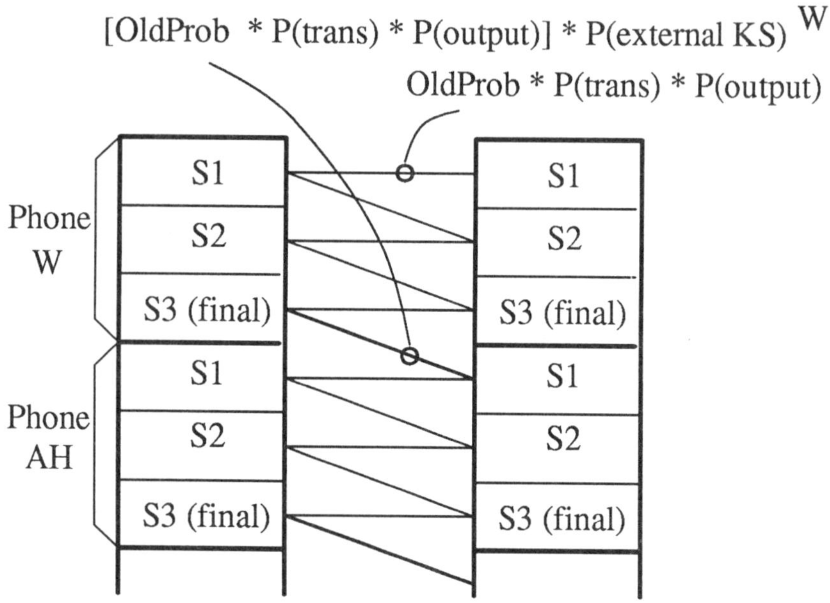

Figure 5-2: Segment-level integration of external variable-width parameters in a Viterbi search. *W* is an empirically determined weight to account for the different dynamic ranges of the probabilities.

where $P_w(d)$ is the probability the word w is d frames long, and μ_w is the expected duration of word w. These parameters were estimated by first using a Viterbi alignment to determine all phone and word boundaries. Then we assumed univariate Gaussian distribution, and estimated the mean and the variance for each phone and word. We now have two distributions for each word, one from occurrences of the word, and the other from the concatenation of the phones. These two distributions were combined together using a weight that depended on the frequency of occurrence of the word. The 51 $P_w(d)$'s are computed from the resulting distribution, and stored for each word. For durations greater than μ_w by more than 25, we used $P_w(\mu_w+25)$, and for durations less than μ_w by more than 25, we used $P_w(\mu_w-25)$. This duration modeling method imposes virtually no overhead.

5.2.2. Knowledge-based Parameters

Knowledge-engineering researchers have argued that many useful speech parameters are segment-based, rather than frame-based. Most of these measurements are more complex, such as duration for aperiodic energy before and after vowel, and frequency location of formants [Cole 83]. These parameters are segment-based, and different parameters are used for different types of sounds. Thus, it is very difficult to integrate them into an HMM-based system.

Instead, we chose to integrate these parameters in the same manner that we integrated word duration. Instead of directly integrating these features, we integrate into SPHINX the *probabilistic phonetic network* of the ANGEL System [Chigier 88], which was derived from these features. This network contains phonetic hypotheses throughout a sentence, with their associated probabilities. To integrate this information into SPHINX, every time a transition leaving a phone is encountered in the search, the begin and end times of that phone are recovered from the Viterbi search, and are used to look up the probability of such a phone in the ANGEL network. If this phone is found in the network, the probability assigned by the network is weighed and combined with the HMM score. Otherwise, some floor probability is used.

ANGEL phone probabilities are integrated into SPHINX exactly the same way as duration probabilities, as shown in Figure 5-2. Due to the different assumptions and nature of the probabilities from SPHINX, duration, and ANGEL network, they have to be weighed before combination. We have experimented with various weights using a set of tuning sentences. The optimal choice was used on the testing sentences. The results are rather disappointing, and will be reported in Section 5.4.

5.3. Lexical/Phonological Improvements

In our first implementation of SPHINX, we assumed that the pronunciation of each word is a concatenated sequence of phones. Similar assumptions were made by CMU's DRAGON [Baker 75b], HEARSAY [Lesser 75], the phonetic version of IBM's TANGORA [Bahl 88b], and BBN's BYBLOS [Chow 87]. While this is the way most of us learned English, it is not a correct assumption. For example, the baseform for the word *Atlantic* is `/ae t l ae n t ix k/`, and the actual pronunciation may be:

```
/ae - t l ae n - t ix - k/
/ae - t l ae n ix - k/
/ae - t l ae n ix -/
/ae - l ae n ix -/
```

If we train the HMMs with the false assumption of one pronunciation per word, the alternate pronunciations will be absorbed into the phone models. We showed that reasonable results can be obtained, but the models have been contaminated by alternate pronunciations. We no longer have purely phonetic models, which may cause problems in making fine phonetic distinctions.

An alternative is to represent a word with a network of pronunciations. A popular network generation method is to begin with a single *baseform* (the standard pronunciation) of the word, and then apply *phonological rules* to incrementally build a network of pronunciations [Cohen 74, Rudnicky 87]. These phonological rules may insert, delete, or substitute phones in a word. Figure 5-3 shows the word pronunciation network for the word *Atlantic* in the dictionary of the CMU ANGEL System.

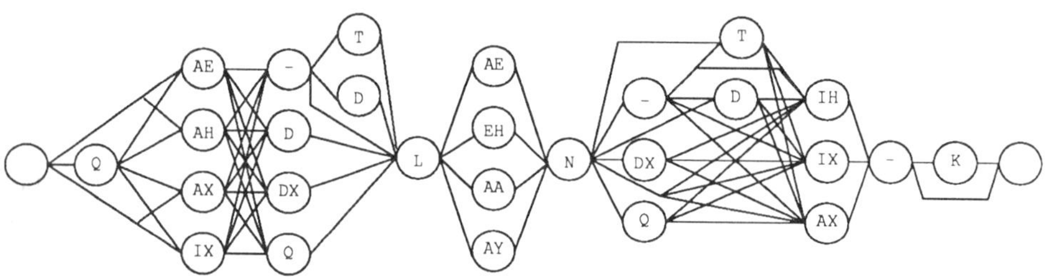

Figure 5-3: Word network for the word *Atlantic* used by the ANGEL System.

Phonetic models trained from networks of pronunciations will be purer than those trained from baseforms. The forward-backward algorithm is not forced into training phones that are not articulated. This advantage, however, is accompanied by many disadvantages. First, training and recognition are both much slower when large pronunciation networks are used. Second, if the Viterbi algorithm is used for decoding, words with more alternate pronunciations are unfairly penalized. Third, how does one determine the probability of each alternate pronunciation?

The third problem is perhaps the most detrimental. Ideally, we would like to estimate these probabilities, but there are too many parameters. For example, the word *Atlantic* in Figure 5-3 has 6912 pronunciations. We could not possibly have enough data to adequately train all these transition probabilities. Even if training data were available, it would contradict the primary motivation for using phonetic models, namely, that new words may be added without training. Another way is to manually assign probabilities (possibly uniform) to all the transitions, and hold them fixed throughout training. But it is well known that, for HMMs, making bad assumptions is better than manually tuning parameters [Bahl 83a].

On the one hand, we need the speed, convenience, and optimality of the baseform approach. On the other hand, we want to use our phonological knowledge to produce a better dictionary. We will now discuss several experiments that attempt to use this knowledge to improve a baseform-like approach.

5.3.1. Insertion/Deletion Modeling

Insertion and deletion rules are more critical than substitution rules for HMM-based recognition. This is because an inserted or a deleted phonetic event will cause very poor acoustic matches. Substitutions are less crucial because they usually represent similar alternate phones. For example, the first syllable of the word *delete* may be `/d iy/`, `/d ih/`, `/d ix/`, or `/d ax/`. These four phones are somewhat similar, and choosing the most likely one (`/d ix/` in this case) will result in reasonable matches during recognition no matter which of the four was actually spoken. As long as we do a good job in choosing the most likely pronunciations for each word, the trained phone models should still be true to their phonetic properties.

The complexity of word networks is largely due to substitution rules, because (1) there are more substitution rules than insertion/deletion rules, and (2) substitutions introduce more states and transitions than deletions or insertions. In view of these facts, we suggest that a pronunciation network using only insertion and deletion rules can benefit from the advantages of the baseform and the network approach.

Not all insertions and deletions are mandatory. In most cases, some phones are optionally deleted or inserted. Moreover, some insertions or deletions may be more likely than others. Therefore, in addition to emphasizing insertions and deletions, we should use a probabilistic model.

There are two ways to model probabilistic insertions and deletions of events. The first method is *explicit insertion/deletion modeling*. The set of phones are divided into two types: (1) mandatory phones, and (2) optional phones. Mandatory phones are those described in the last chapter, and must consume at least one frame of time. Optional phones have the same structure as mandatory phones, except an additional null transition is added between the initial state and the final state. This null transition allows the phone to be skipped completely with probability specified by the transition probability of the null transition. Unlike the network approach, these deletion probabilities are easily trainable since they are not word-dependent, but phone-dependent. The penalty we pay for trainability is that the probability of skipping a phone is not sensitive to context. One way to overcome this is to use context-dependent insertion/deletion probabilities, but we have not experimented with them since we may not have sufficient data to train these probabilities.

Another weakness of this approach is that modeling of deletion of single phones is not sufficient. For example, the word *Atlantic* in Figure 5-4 has three pairs of deletable phones. In the first pair, only one of /-/ and /t/ may be deleted, while in the second pair, both /-/ and /t/ may be deleted. The first case cannot be modeled using context-independent explicit insertion/deletion modeling.

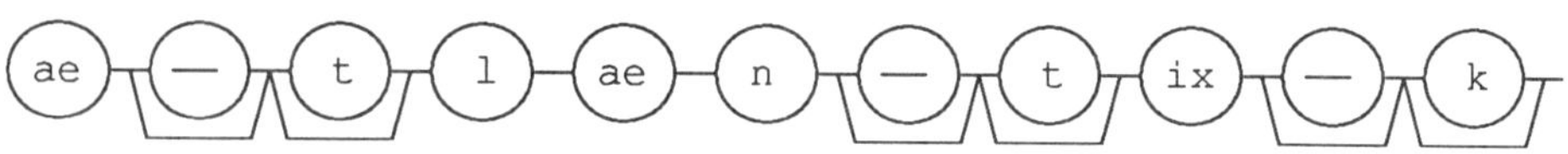

Figure 5-4: The baseform for *Atlantic* using explicit insertion/deletion modeling. Null transitions are added to indicate that the phone may be skipped. The probability of the skip is dependent on the phone, not on the word.

Alternatively, insertions and deletions can be *implicitly modeled*. We create *compound phones* that encompass the inserted or the deleted events, so that the insertion/deletion probabilities are learned implicitly, in both transition and output parameters of the models. For example, to model released stops and closures, such as ***beam***, ***ten***, and ***April***, we use one type of compound phone. For these phones, the stop is always present, but the closure is occasionally missing. This phenomenon can be modeled within the HMM. For stops that need not be released, such as ***deleted*, *Hepburn*,** and *Bud-Test*, we have another set of compound phones, /dd/, /pd/, /td/, /kd/. For these phones, it is possible to see (1) only the closure, (2) closure and the stop, and (3) only the stop, in that order of likelihood. We simply model all of these variations with one HMM, and let the forward-backward algorithm learn their likelihoods. An example of *implicit insertion/deletion modeling* is shown in Figure 5-5.

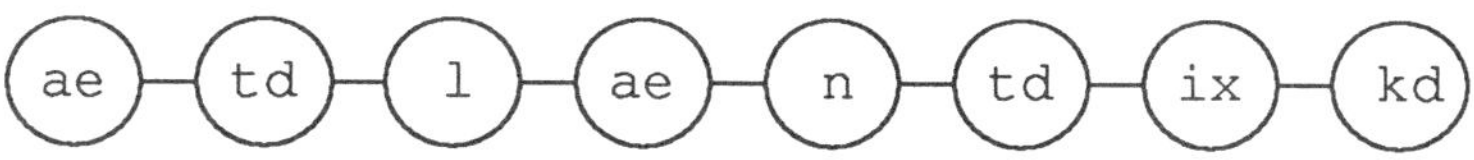

Figure 5-5: The baseform for *Atlantic* using implicit insertion/deletion modeling. Compound units such as /td/ and /kd/ are defined. The deletion of closure or burst within those units are modeled by the HMM parameters.

To show that the two types of compound phones are significantly different, the HMMs representing /t/ and /td/, with transition probabilities, are depicted in Figure 5-6. /td/ has a much higher probability (0.49) than /t/ (0.14) of being two or three frames long. Furthermore, when they are both four frames or longer, /td/ has the same closure (pdf "B") duration as /t/, but has a much shorter burst (pdf "M" and "E") duration.

Compared to explicit insertion/deletion modeling, implicit modeling has the advantage of trainability (fewer models) and flexibility (with adjacent deletable events). It should work well for events like closure-stop pairs because they almost always appear together. But rarer pairs cannot be modeled this way. Therefore, only stop-closure pairs are modeled using implicit insertion/deletion modeling.

5.3.2. Multiple Pronunciations

So far, we have ignored the fact that some substitutions may result in drastically different pronunciations. Some words have multiple pronunciations, such as *the* (/dh ax/ and /dh iy/). Other words may have substitution rules that result in substantially different realizations, such as *for* (/f ao r/ and /f er/). Using our current dictionary, which does not permit networks of pronunciations, the only way to have multiple pronunciations is to add multiple entries of the same word in the dictionary.

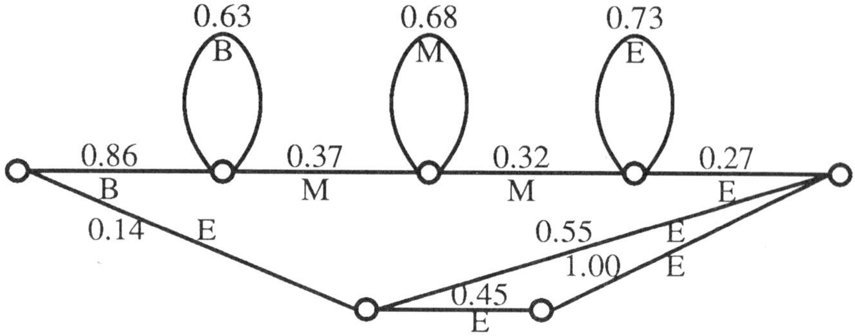

(A) HMM for /t/

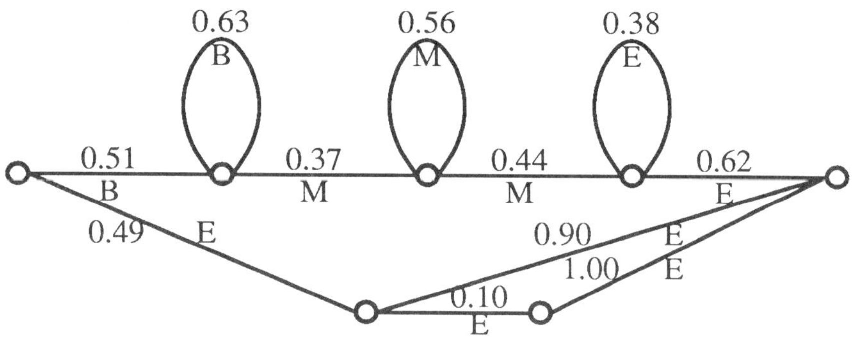

(A) HMM for /td/

Figure 5-6: HMMs with transition probabilities for the phones /t/, which consists of an optional but likely closure and a released t, and /td/, which consists of an optional but likely closure, and a t which is usually not released.

During training, all alternate pronunciations of a word are connected in parallel and are trained together. We have implemented multiple-pronunciation training and recognition algorithms; however, they did not lead to any improvements [Hwang 88]. Adding multiple pronunciations usually saved some instances when an infrequent pronunciation of a word was used (such as /ey/ for *a*), but also caused new confusions, resulting in little or no improvement overall. Thus, every version of SPHINX described in this monograph assumes one pronunciation per word.

5.3.3. Other Dictionary/Phone-Set Improvements

This section describes a number of additional experiments we performed with the dictionary and the phone set.

5.3.3.1. Phonological Rules

In order to improve the appropriateness of the word pronunciation dictionary, a small set of rules were adopted from the ANGEL Word Group, and modified to suit our needs [Polifroni 88].

A set of rules was implemented to modify closure-stop pairs into optional compound phones (stops that may not be released), as described in the previous section. Of the six stops, `/d/`, `/p/`, `/t/`, `/k/` have two models depending on whether the stop is expected to be released. The other two stops, `/b/` and `/g/`, almost always appeared as released stops in the resource management task.

Another set of rules governed flapping of `/t/`'s and `/d/`'s into `/dx/`. Flapping is a very common practice in American English. For example, almost all American English speakers would say `/b eh dx er/` rather than `/b eh t er/` for the word *better*. Therefore, it is reasonable to replace all the appropriate `/t/`'s and `/d/`'s with `/dx/`'s.

Another rule was applied to reduce nasal `/t/`'s. For example, the `/t/` in *twenty* is usually not articulated, and this rule deletes it. While this rule may appear dangerous, experimental results showed that it produced slightly better performance. A final rule folded `/zh/` to `/sh/` because the affricate `/zh/` was very infrequent in our task.

5.3.3.2. Non-Phonemic Affricates

Although the phonemes in English are well-defined, there are actually many sounds which are frequently used, but are not phonemic. For example, stop-fricative pairs such as `/ks/`, `/ps/`, `/ts/`, `/bz/`, `/dz/`, or `/gz/`, are actually quite different from the concatenated phoneme pairs. They appear more like different affricates. Thus, it is sensible to model them as separate phones; however, experiments showed that only `/ts/` improved the recognition rate for our task. This is because it improves discrimination among `/s/`, `/t/`, and `/ts/`, which is important in our task, where the first word in a sentence is often *what*, *what's*, or *was*. In addition, there are possessive and plural forms of many words that end with `/t/`, so modeling

/ts/ as a separate unit improves the discrimination among these words, and is trainable because of its frequency of occurrence. Therefore, in the standard SPHINX dictionary, we model /ts/ as a separate unit.

Although we have not obtained any improvements from modeling the other non-phonemic affricates, we believe that it is sensible to do so for a larger task. There are also other phoneme pairs that could be modeled as compound units, such as /dr/ and /tr/. In Chapter 6, we will discuss modeling phones in context; however, the distortion to the phones in the phone pairs we discussed may be too large, and modeling them as compound units would lead to superior results if we could adequately train them.

5.3.3.3. Tailoring HMM Topology

Finally, there is the issue of what HMM topology is optimal for phones in general, and what topology is optimal for each phone. We have experimented earlier with a variety of topologies for phone recognition, and found that although the choice of model was not critical for continuous speech recognition, our model (as shown in Figure 4-2) led to better results than the other models we tried.

For word recognition, we first tried different topologies for different classes of phones. For example, silence and fricatives may require fewer states and distributions, while diphthongs may require more. However, this led to a slight degradation in result.

We also experimented with different labelings of the five lower transitions shown in Figure 5-7, holding the topology fixed. In other words, if a phone could be one, two, or three frame long, which of the three sets of output pdf's is the most appropriate? The other transition labels were fixed, because they are temporally constrained. Since there were many possible labelings, and many phones to test, it was not practical to use recognition results as a criterion for this experiment. Instead, we used hand-labeled phones from the TIMIT database, and trained every phone HMM with every possible transition labeling. The labeling that yielded the highest probability of producing the training data for a phone was selected. Table 5-2 shows the labeling used for each phone.

Finally, although each phone model contains five transitions and two states to model durations of 1, 2, and 3 frames, not all phones could be so short. Therefore, an HMM reduction procedure was applied to remove all

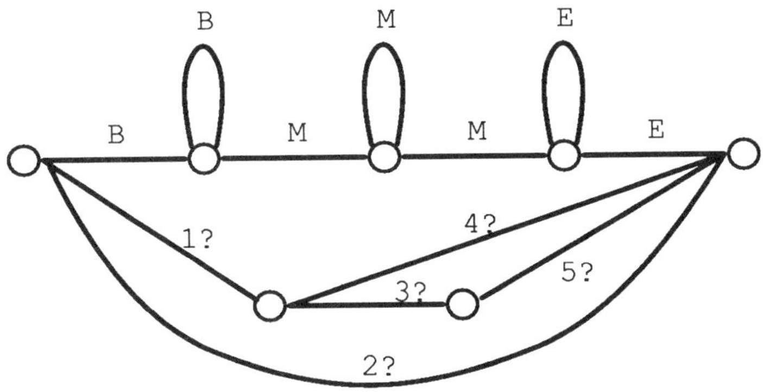

Figure 5-7: The HMM topology used in SPHINX, with different output pdf labelings on the lower transitions for different phones.

1	2	3	4	5	Phones
B	B	B	B	E	/ae/, /eh/, /ah/, /aa/, /ao/, /uw/, /aw/, /ay/, /ey/, /ow/, /oy/, /l/, /en/, /er/, /m/, /n/, /ng/, /f/, /sh/, /v/, /z/, /sil/
B	B	M	E	E	/ih/, /iy/, /uh/, /ax/, /ix/, /r/, /w/, /y/, /ch/, /jh/, /dx/
E	E	E	E	E	/b/, /d/, /dh/, /g/, /k/, /p/, /t/, /s/, /th/, /hh/, /ts/

Table 5-2: Lower transition labels assigned for each phone using the HMM in Figure 5-7.

transitions with probability of zero, and then remove all states (except the initial state) that have no incoming transitions, and all states (except the final state) that have no outgoing transitions. This procedure is iterated until an iteration where nothing was removed. This resulted in about 10% reduction in the number of transitions, and 5% reduction in the number of states for the Viterbi search.

5.3.3.4. Final Phone Set and Dictionary

Experimenting with the phone set and dictionary improvements described in this section, we found that the best results were obtained with implicit insertion/deletion modeling, compound phones for /-d/, /-p/, /-t/, /-k/, and /ts/, the phonological rules described above, and improved HMM models. The results will be discussed in the next section.

Table 5-3 lists the set of phones, and Table 5-4 shows a section of our final phonetic pronunciation dictionary.

Phone	Example	Phone	Example	Phone	Example
/iy/	*beat*	/l/	*led*	/t/	*tot*
/ih/	*bit*	/r/	*red*	/k/	*kick*
/eh/	*bet*	/y/	*yet*	/z/	*zoo*
/ae/	*bat*	/w/	*wet*	/v/	*very*
/ix/	*roses*	/er/	*bird*	/f/	*fief*
/ax/	*the*	/en/	*button*	/th/	*thief*
/ah/	*bmtt*	/m/	*mom*	/s/	*sis*
/uw/	*boot*	/n/	*non*	/sh/	*shoe*
/uh/	*book*	/ng/	*sing*	/hh/	*hay*
/ao/	*bought*	/ch/	*church*	/sil/	(silence)
/aa/	*cot*	/jh/	*judge*	/dd/	*deleted*
/ey/	*bait*	/dh/	*they*	/pd/	*ship*
/ay/	*bite*	/b/	*bob*	/td/	*set*
/oy/	*boy*	/d/	*dad*	/kd/	*comic*
/aw/	*bough*	/g/	*gag*	/dx/	*butter*
/ow/	*boat*	/p/	*pop*	/ts/	*its*

Table 5-3: List of the improved set of phones in SPHINX.

5.4. Results and Discussion

The results of various versions of SPHINX using ideas presented in this chapter are shown in Table 5-5. The version abbreviations are defined in Table 5-6. These versions were incrementally built and tested. If an idea was found to be helpful, it was used for all subsequent experiments. Such ideas are ***italicized***.

Our results with bilinear transform corroborated with the earlier results that advocated mel-scale coefficients [Zwicker 61, Davis 80, Shikano 86c]. We found that bilinear transformed coefficients yielded a significantly higher recognition rate. An even greater improvement came from the use of

Word	Baseform	After rules
ADDED	/ae d ix d/	/ae dx ix dd/
ADDING	/ae d ix ng/	/ae dx ix ng/
AFFECT	/ax f eh k t/	/ax f eh k td/
AFTER	/ae f t er/	/ae f t er/
AGAIN	/ax g eh n/	/ax g eh n/
AJAX	/ey jh ae k s/	/ey jh ae k s/
ALASKA	/ax l ae s k ax/	/ax l ae s k ax/
ALERT	/ax l er t/	/ax l er td/
ALERTS	/ax l er t s/	/ax l er ts/

Table 5-4: A section of the SPHINX dictionary with word, original baseform, and the pronunciation after rule application.

Version	No grammar	Word Pair	Bigram
Baseline	31.1% (25.8%)	61.8% (58.1%)	76.1% (74.8%)
BT	34.2% (28.6%)	63.1% (59.4%)	78.5% (76.0%)
4F1C	41.5% (36.0%)	78.7% (76.2%)	86.2% (84.8%)
4F3C	45.6% (40.1%)	83.3% (81.1%)	88.8% (87.9%)
Exp.	46.8% (42.7%)	85.1% (82.4%)	89.3% (88.5%)
Imp.	50.0% (45.3%)	86.8% (84.4%)	91.2% (90.6%)
ANGEL-PPN	49.8% (44.8%)	--------- --------	--------- ---------
Dur.	55.1% (49.6%)	85.7% (83.8%)	91.4% (90.6%)

Table 5-5: The SPHINX results with knowledge enhancements. ***Italicized*** enhancements are used for all subsequent experiments.

differential coefficients and power. When all 26 coefficients were combined in one codebook using a composite distance metric, the results improved drastically. When we separated these feature sets into three codebooks, even better results emerged. We did not study the contribution of each feature set

Version	Description
Baseline	The version used at the end of last chapter.
BT	Adding bilinear transform ***(used in all subsequent versions).***
4F1C	Using four features sets (cepstrum, differenced cepstrum, power, and differenced power) in one codebook.
4F3C	Using four feature sets in three codebooks ***(used in all subsequent versions).***
Exp.	All the dictionary and phonological improvements, plus explicit insertion/deletion modeling.
Imp.	All the dictionary and phonological improvements, plus implicit insertion/deletion modeling ***(used in all subsequent versions).***
Dur.	Segment-level integration of word duration probabilities ***(used in all subsequent versions without grammar).***
ANGEL-PPN	Segment-level integration of ANGEL's probabilistic phonetic network.

Table 5-6: The definition of the version abbreviations used in Table 5-5. ***Italicized*** versions are used in all subsequent experiments.

and additional codebook for word recognition; however, our phoneme recognition experiments showed that each feature set and each codebook contributed significantly [Lee 88b].

Next, we improved the dictionary and the phone set, as described in Section 5.3. We found that explicit modeling of optional phones led to a much smaller improvement than did implicit modeling. By looking at the tuning results, we saw that with explicit modeling, SPHINX sometimes took double-skips on optional phone pairs (such as closure-stop), which was detrimental. Another problem was the modeling of optional events without taking context into consideration. Moreover, we saw earlier that implicit modeling produced distinct models for `/t/` and `/td/`, which was effective. Finally, implicit modeling did not require null transitions within a phone, which improved the speed of recognition considerably. Therefore, we abandoned the idea of explicit insertion/deletion modeling and the use of within-phone null transitions, and will use implicit insertion/deletion modeling with compound models in all subsequent experiments.

We were unable to obtain any improvement by integrating the ANGEL network into SPHINX. This is probably due to several reasons:

1. ANGEL is less accurate than SPHINX even for phoneme recognition. ANGEL's phonetic accuracy is about 55%, while SPHINX's is about 74% [Lee 88b].
2. The behavior of the ANGEL phone hypothesizer was usually reasonable, but occasionally it would completely miss events that are fairly obvious. This resulted in zero (actually floor) probabilities which was detrimental to our probability multiplication scheme.
3. The ANGEL phonetic network was designed to recognize pure phones, while SPHINX absorbed many imperfect assumptions about phones; therefore, SPHINX was able to "pretend" that a phone was there while ANGEL was not.

We expect that if ANGEL could be retrained with SPHINX-segmented speech rather than hand-labeled speech, it could improve the recognition rate of SPHINX. But a significant improvement would require a major enhancement of the ANGEL acoustic/phonetic accuracy.

To illustrate the importance of error modeling, we trained a set of phonetic HMMs using hand-labeled speech from the TIMIT database. We used these HMMs for recognition of the resource management database with two dictionaries: (1) one pronunciation per word as before, and (2) the dictionary used by ANGEL, with a network of many pronunciations per word. The results are shown in Table 5-7. While these results are extremely poor, they illustrate several important lessons. Although our phonetic HMMs are now "pure" (since they were trained with carefully hand-labeled speech), the results are considerably worse. This is because the TIRM-trained HMMs absorbed and modeled the errors in the pronunciation dictionary. It is also interesting to note that, for TIMIT-trained HMMs, pronunciation network led to much worse results than that with single pronunciation. Although adding alternate pronunciations provided a right path for any pronunciation of a word, it also provided hundreds or thousands of potential mismatches for other words. Moreover, the large network necessitated extensive pruning, which led to some search errors. As a result, attempting to correct for system problems by hand led to much worse results.

Training Data	Lexical Representation	Recognition Accuracy
TIRM	Single pronunciation	50.0% (45.3%)
TIMIT	Single pronunciation	38.5% (32.1%)
TIMIT	Pronunciation Network	24.1% (14.8%)

Table 5-7: Word recognition results (no grammar) using TIMIT-trained pure phonetic models vs. TIRM-trained models, which absorbed the erroneous assumptions of the phonetic dictionary.

Finally, the addition of duration information significantly improved SPHINX's accuracy when no grammar was used, but was not helpful with a grammar. For the word-pair grammar, duration modeling actually degraded the accuracy because we used a tuning set to tune the weights, which did not work well on the test set. When no grammar is used, many more hypotheses are considered, and duration information can be used to filter out many hypotheses with implausible word durations. When a grammar is used, much more constraint is applied, which sharply decreases the utility of duration. Also, the use of a language model and duration model requires the tuning of two weights, which we may not have optimized. Therefore, in the future versions, duration modeling will be used only when no grammar is used.

5.5. Summary

In this chapter, we attempted to improve the performance of SPHINX by enhancing it with human knowledge. To that end, we have been successful; however, it may be surprising to some that the greatest improvements came not from intricate acoustic-phonetic parameters or elaborate phonological rules, but from crude knowledge of speech and English.

Examples of such speech knowledge include the use of bilinear transform for frequency warping, which is in agreement with studies in human perception. We also used differential coefficients, which corroborates with spectrogram readers who find relative formant information more significant than absolute ones. Finally, power and differential power are important prosodic cues. These are all examples of simple human speech knowledge that could be incorporated in an HMM recognizer as *fixed-width* parameters. We investigated different approaches to integrate fixed-width parameters using discrete HMMs, and found the *multiple-codebook* approach to be superior. These new features and codebooks substantially improved our accuracy. They reduced the error rate of the word-pair grammar by 43%, and the error rate of the bigram grammar by 52%.

We then questioned the appropriateness of our dictionary and phone set. We combined stop-closure pairs into compound models, and used two representation for each such pair to separately model closure-stop pairs with mandatory stops, and those with optional ones. We found that this approach, *implicit* insertion/deletion modeling, led to better results than alternative approaches. We also found that it was helpful to merge other stop-fricative pairs, since the acoustic nature of these pairs are significantly different from their concatenation. In addition, we used several rules to capture simple phonological effects that did not deviate from our one pronunciation per word assumption. Finally, we used different HMM topologies for different phones. These optimizations resulted in modest, but significant, improvements in recognition accuracy.

We experimented with two types of variable-width parameters. We were unable to obtain any improvements from the addition of the probabilistic phonetic network derived from more complex features. This is mostly due to the inadequacy of the ANGEL hypotheses, and the incompatibility of ANGEL's pure phonetic hypothesization. However, we did obtain a non-trivial improvement from word-duration modeling when no grammar was used.

From the experiments in this chapter, we found that it is possible to substantially improve an HMM system if we introduce knowledge that is compatible within the HMM framework. We also found that although HMMs are good at absorbing inaccurate but consistent assumptions, it still may be fruitful to replace these assumptions with more accurate ones. Finally, we found that it is possible to add new knowledge to HMM recognizers without training, but the quality of these new knowledge sources is crucial.

6 Finding a Good Unit of Speech

Given that we will use hidden Markov models to model speech, one important question is: what unit of speech should an HMM represent? In the previous chapters, we have used phones as the fundamental unit of speech. An even more natural unit is words. In this chapter, we will discuss the strengths and weaknesses of word and phone models, as well as a number of other units proposed by earlier work. Then, we will propose two new units that will substantially improve the performance of speaker-independent continuous speech recognizers. Finally, we will present comparative results of different variations of these units.

6.1. Previously Proposed Units of Speech

6.1.1. Words

Words are the most natural units of speech because they are exactly what we want to recognize. Also, word models are able to capture within-word contextual effects. For example, the phone `/t/` in *ten* is as expected, the phone `/t/` in *thirty* is usually flapped, and the phone `/t/` in *twenty* may be deleted. By modeling words as units, these phonological variations will be assimilated. Therefore, when word models can be adequately trained, they will usually yield the best performance. This is demonstrated by the success of several recent small-vocabulary word-based recognizers [Lippmann 87, Rabiner 88a].

However, using word models in large-vocabulary recognition

introduces several grave problems. Since training data cannot be shared between words, each word has to be trained individually. In order to train word models adequately, considerable training data (probably 20 or so in our system) are needed. But for a large-vocabulary task, this imposes too great a demand for training data. This problem is difficult for speaker-independent systems, and even more difficult for speaker-dependent ones.

Another problem is that memory usage grows linearly with the number of words, since there is no sharing between words. For a 20,000-word task, this will require on the order of a gigabyte of main memory.

Finally, for many tasks, it would be convenient to provide the user with the option of adding new words to the vocabulary. If word models are used, the user would have to produce many repetitions of the word, which is extremely inconvenient. For speaker-independent recognition, repetitions from *different* speakers are needed, which is even more troublesome.

Therefore, while word models are natural and model contexts well, because of the lack of sharing across words, they are not practical for large-vocabulary speech recognition.

6.1.2. Phones

In order to allow sharing across words, some subword unit has to be used. The subword units most familiar to us are the phones of English. The implementation of SPHINX we have described thus far is based on phone models.

Unlike word models, phone models are highly trainable. Since there are only about 50 phonemes in English, they can be sufficiently trained with just a few hundred sentences. They are also task-independent, and can be trained on one task and tested on another.[14] Finally, phones are well-understood, and most English speakers can, or can be trained to, produce the most likely phone sequence for any word.

We have seen that the earlier implementations of SPHINX yielded reasonably accurate results. Other work that used phone models include

---

[14]Although performance will likely deteriorate due to task-dependent contexts implicitly modeled by context-independent phone models.

[Bahl 80a, Merialdo 87, Lee 87, Lee 88a, Murveit 88]. While these results are satisfactory, well-trained word models will always outperform well-trained phone models due to the aforementioned advantages of word modeling. This is supported by work at IBM [Bahl 88b], which showed that word-based DTW performed significantly better than phone-based HMM for speaker-dependent recognition, as well as by Lincoln Labs [Paul 88], which showed that word-based HMM resulted in 50% error rate reduction from phone-based HMMs.

The reason that phone models are inadequate is: they assume a phone in any context is equivalent to the same phone in any other context. However, this is far from the truth [Reddy 83]. Although we may try to say each word as a concatenated sequence of phones, these phones are not produced independently, because our articulators cannot move instantaneously from one position to another. Thus, the realization of a phone is strongly affected by its immediate neighboring phones. Figure 6-1 illustrates co-articulatory effects on the phoneme `/t/` in four different contexts.

Another problem with using phone models is that phones in function words, such as *a*, *the*, *in*, *me*, are often articulated poorly, and are not representative instances of the phones.

While word models are poor units because they lack generality, phone models are also poor because they over-generalize.

6.1.3. Multi-Phone Units

One way to model co-articulatory effects is to use larger units of speech. Examples of this include syllables [Hunt 80], or demisyllables [Rosenberg 83]. These units encompass the phone clusters which contain the most severe contextual effects. However, while the central portions of these units have no contextual dependencies, the beginning and ending portions are still susceptible to some contextual effects.

A more serious problem is the large number of these units. For example, there are over 20,000 syllables and over 1000 demisyllables in English. Although this may be a reduction from word models in a very large vocabulary, there are still too many parameters to reliably estimate when different units cannot share the same training data. Finally, experiments

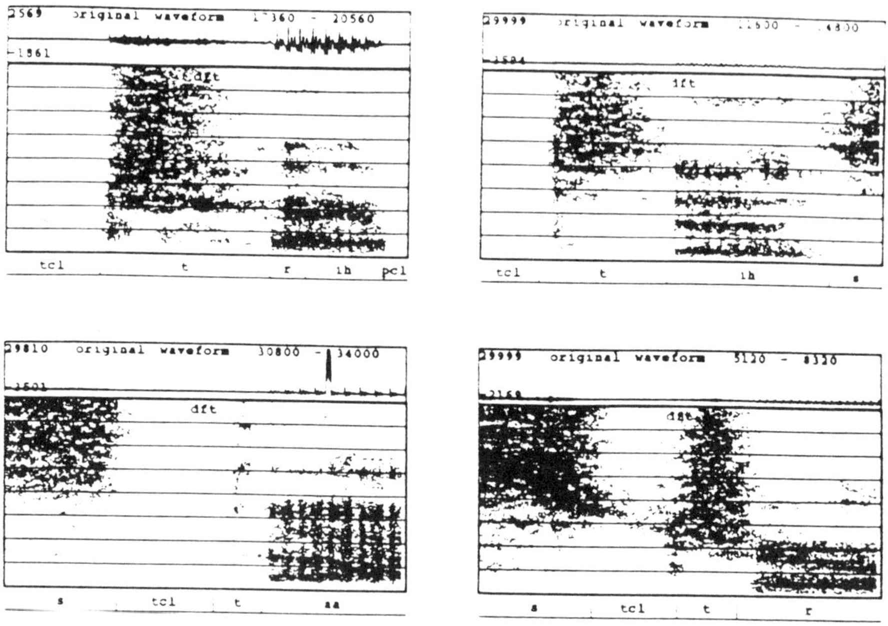

Figure 6-1: The waveforms and spectrograms for the phoneme /t/ in four different contexts: part of /t r/, left of /ih/, part of /s t/, and part of /s t r/. It is clear that the realization of /t/ is highly dependent on context, and that a context-independent /t/ model is inadequate.

[Rosenberg 83] showed that a demisyllable-based recognizer performed substantially worse than a word-based recognizer.

6.1.4. Explicit Transition Modeling

Since transitions in and out of phones are poorly modeled by phone models, one solution is to model these transitional regions explicitly. Diphones [Schwartz 80, Klatt 86] model pairs of phones without the use of stationary phones. Another approach is to use stationary phone models and insert transition models [Cravero 86].

Transition modeling suffers from the same problem as multi-phone units. Instead of N phones, there are N^2 phone transitions. Like multi-phone units, these units cannot easily share training data. Therefore, transition models also result in too many parameters to estimate when there is no sharing.

6.1.5. Word-Dependent Phones

Word-dependent phones are a compromise between word modeling and phone modeling. Although phone models are used, these phone models are *word-dependent*. In other words, a phone model used for one word has different parameters from the model of the same phone in another word. Like word models, word-dependent phone models can model word-dependent phonological variations, but require considerable training and storage.

However, word-dependent phone modeling is better than word modeling in two ways: First, when a word has not been observed frequently, its parameters can be interpolated (or averaged) with that of context-independent phone models. This is what we meant earlier by "sharing". Second, a new word need not be repeated many times—context-independent phone models can be used to get acceptable performance. The same is true for a word in the vocabulary but not observed in the training data.

Word-dependent phone modeling was proposed by Chow *et al.* [Chow 86]. In that study, word-dependent phone models were interpolated with context-independent phone models using empirically determined weights. Word-dependent phone modeling yielded a 10% error rate, while word models had a 14% error rate, and phone models had a 24% error rate. Word-dependent phone models actually outperformed word models, because some word models were poorly trained while the corresponding word-dependent phone models were reasonably trained through interpolation. Murveit and Weintraub [Murveit 88] also used word-dependent phone modeling; however, their improvement was considerably smaller. We believe that this is because they did not interpolate the word-dependent phone models with better trained, but less appropriate, context-independent phone models.

6.1.6. Triphones (Context-Dependent Phones)

Context-dependent phone models are similar to word-dependent phone models; instead of modeling phone-in-word, they model phone-in-context. A context usually refers to the immediate left and/or right neighboring phones. A *left-context dependent phone* is dependent on the left context, while a *right-context dependent phone* is dependent on the right context. A *triphone* model takes into consideration both the left and the right neighboring phones; if two phones have the same identity but different left or right context, they are considered different triphones. Triphone models are usually poorly trained because there are many triphones. But since triphone models are specific phone models, they can be interpolated with better-trained but less appropriate context-independent models.

Bahl *et al.* [Bahl 80a] first proposed context-dependent modeling of phonemes. Schwartz *et al.* [Schwartz 84] at BBN were the first to publish comparative results of triphone modeling. In that study and later studies [Schwartz 85, Chow 86], triphone models were interpolated with right-context-dependent models (phone models that are dependent on the right context), left-context-dependent models, and context-independent models. Each pdf in each model was given a different weight according to appropriateness (for example, left-context models have greater weights for leftmost pdf) and amount of training (for example, if a triphone has been observed many times, its weight will dominate). This weight matrix was tuned by hand. For both phoneme and word recognition, modeling phone-in-context reduced the error rate by about 50%.

Triphone modeling is powerful because it models the most important co-articulatory effect, and is much more sensitive than phone modeling. To a large extent, word-dependent phones are really modeling phone contexts because most words do not have truly word-dependent effects that cannot also be predicted from context. Another advantage over word-dependent phones is that context-dependent phone models are relatively task-independent. Given a new task, it is not strictly necessary to retrain with sentences from that task, whereas with word-dependent phone modeling, retraining is necessary if the new vocabulary is not a subset of the old one.

While triphone modeling is a powerful idea, it has two problems. The first problem is memory wastage. When a triphone is observed once, a model is created for it, and with a large number of triphones, the memory used could be substantial. For example, with the 1000-word task, there are 2381 triphone contexts, which requires 24 megabytes of memory in our system.[15] For natural English, this number is much larger. While this memory problem can be overcome, the other problem is more serious:

[15]Each phone HMM takes about 10K of memory.

triphone modeling assumes that every triphone context is different. Actually, many phones have similar effects on other phones. For example, /b/ and /p/ are both labial stops, and have similar effects on the following vowel. It would be desirable to find instances of similar contexts and merge them together. This would lead to reduced memory usage and better-trained models.

6.1.7. Summary of Previous Units

In the preceding sections, we have evaluated several previously proposed units of speech. We emphasized three important properties for speech units, namely, sensitivity, trainability, and sharability. A unit is sensitive if it accounts for co-articulatory effects. If we have infinite training data, sensitivity is the only property of interest. But because our training data is not only finite, but often limited, trainability becomes an important issue. Trainability can be achieved by using very general units at the cost of insensitivity, or by sharing among units.

Table 6-1 evaluates the appropriateness of the units we described for large-vocabulary recognition using these three criteria. The word is a sensitive unit, but it is not easily trainable nor sharable. Phones are insensitive, but they are so trainable that sharing is not needed. Multi-phone units and transition units are sensitive, but are not easily trainable and provide no means of sharing. Word-dependent phones and context-dependent phones are sensitive, and can be trained because there exist means of sharing. Therefore, both are very appealing units. Our only criticism is that sensitivity is achieved with too fine a level of detail. More generalization can lead to fewer models and a higher level of trainability.

6.2. Deleted Interpolation of Contextual Models

Throughout the preceding sections, we have emphasized the importance of detailed models, and of using robust models to smooth detailed ones. We have also cited empirical results that substantiate our claim. Now we will address the question of how to smooth two distributions representing the same event, but were trained with varying degrees of robustness and detail.

BBN has successfully used hand-tuning of a weight matrix that took

Unit	Sensitivity	Trainability	Sharability
Word model	Yes	No	No
Phone model	No	Yes	Not needed
Multi-phone model	Yes	Difficult	No
Transition model	Yes	Difficult	No
Word-dep. phone model	Yes	Difficult	Yes
Context-dep. phone model	Yes	Difficult	Yes

Table 6-1: Suitability of previously proposed units of speech to large vocabulary recognition.

into consideration for each distribution [Schwartz 84, Schwartz 85]: (1) the amount of training, and (2) the "appropriateness" of the distribution. Although hand-tuning has led to very respectable results for BBN, we believe that hand-tuning is too subjective. It is not clear we could estimate these parameters reliably, and it takes too much time to globally optimize this matrix because there are many free parameters. In this section, we will discuss an automated procedure that determines weights for different distributions.

An ideal solution for weighing different estimates of the same event is *deleted interpolated estimation* [Jelinek 80]. Deleted interpolation weighs each distribution according to its ability to predict unseen data. If we trained two output pdf estimates, b^1_{ij} and b^2_{ij}, derived from two different estimates (say, context-dependent and context-independent), for the transition from state i to state j, and we would like to combine them into:

$$b_{ij} = \lambda_1 b^1_{ij} + (1 - \lambda_1)\, b^2_{ij} \qquad (1)$$

This problem could be viewed as an HMM problem, where the transition with output pdf b_{ij} is replaced by two parallel transitions b^1_{ij} and b^2_{ij}, whose transition probabilities are λ_1 and $(1-\lambda_1)$, respectively. This is graphically illustrated in Figure 6-2. This interpretation of interpolation as an HMM problem immediately suggests the use of the forward-backward algorithm to estimate λ_1.

However, since λ_1 serves to predict *unseen* data, it should be estimated

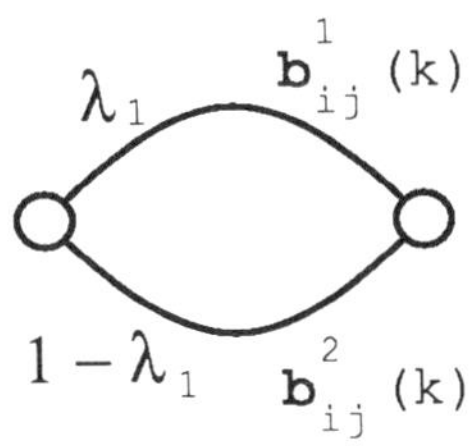

Figure 6-2: Interpolation of two distributions formulated as an HMM problem. b's are two sets of output pdf's, and λ_1 is the transition probability or weight for b^1_{ij}.

from data that was not used to determine b^1_{ij} and b^2_{ij}. In order to do this, we divide our training into several blocks, and use all the blocks except a *deleted* block to estimate λ_1 on that one block. λs are estimated after all possible deletions. The maximum likelihood estimates for λ_1 is simply:

$$\overline{\lambda}_1 = \frac{1}{N}\sum_{i=1}^{N}\frac{\lambda_1 \cdot P_1(y_i)}{\lambda_1 \cdot P_1(y_i) + (1-\lambda_1)\cdot P_2(y_i)} \tag{2}$$

where $P_1(y_i)$ is the probability of producing all the data in block i using distribution 1, which was trained from all N blocks except block i.

The above formulation assumes that the same λ is used for estimate 1 everywhere. In practice, it may be desirable to use a different λ for every phone, or a different λ for every distribution, or we could estimate a different λ for each predetermined range of counts. Furthermore, with the above re-estimation formula, each iteration of deleted interpolation is as expensive as an iteration of the normal forward-backward algorithm. To reduce computation, we could simply keep separate counts for each block during the final iteration of forward-backward, and carry out deleted interpolation on the *counts*, rather than retraining on all the sentences.

In our implementation for training detailed (word or context dependent) models, we first initialized the detailed models with the general (context-independent) models. Two iterations of the normal forward-backward algorithm were run using detailed modeling. During the second iteration, we divided the data into two blocks, and maintained separate output and transition counts for each block. Then, 100 iterations of deleted interpolation were run to combine several estimates. Typically these estimates were:

- A detailed model (word-dependent or context-dependent).
- A general model (context-independent phone models—the counts for a general model are the sum of the counts in all the detailed models that correspond to the general model).
- A uniform distribution.

Thus, this procedure not only combines detailed (but less robust) models with robust (but less detailed) models, but also smooths the distribution using the uniform distribution.

Both output and transition probabilities were interpolated. For output probabilities, a λ was maintained for each context-dependent distribution. For transition probabilities, a λ was maintained for each context-dependent phone. We have experimented with computing a λ for each range of counts, but found distribution-dependent and phone-dependent λs to be better.

Now that we are equipped with the means of combining different estimates for the same events, we will introduce two new types of context-dependent estimates.

6.3. Function-Word-Dependent Phones

Function words, such as *the*, *a*, *in*, *with*, are typically prepositions, conjunctions, pronouns, and short verbs. These words are particularly problematic in continuous speech recognition. Waibel [Waibel 86] showed that in continuous speech only 14% of the function words are stressed, while 93% of the content words are stressed. Unstressed syllables are much harder to recognize [Klatt 72, Lea 80]. While function words are spoken clearly in isolated-word speech, they are articulated extremely poorly in continuous speech. The phones in function words are distorted in many ways. They may be shortened, omitted, or seriously affected by neighboring contexts. For example, Table 6-2 enumerates 50 phonetic transcription labels assigned by CMU expert spectrogram readers for the word *the*. Many other function words have a large number of pronunciations. Since these effects are specific to the individual function words, explicit modeling of phones in these function words should lead to a much better representation.

Function words have caused considerable problems in SPHINX. Among the 684 errors in our system when no grammar was used, 334 were function word errors. Function words take up only 4% of the vocabulary, or about

/dh ax/	/dh ix/	/- dh ax/	/- dh ix/
/ax/	/- dh iy/	/dh ah/	/ix/
/dh ih/	/- d ix/	/iy/	/th ix/
/th iy/	/- d ax/	/- d ih/	/- d iy/
/dh/	/- dh eh/	/d ix/	/dh ao/
/dx ax/	/ih/	/- d ah/	/- dh ao/
/- t ah/	/- t ih/	/- th iy/	/ah/
/d iy/	/dh ax q/	/dh er/	/dh iy ih/
/dh m/	/dx ah/	/dx ih/	/dx ix/
/nx ah/	/nx ey/	/nx ix/	/th eh/
/dh iy/	/th ax/	/- dh ah/	/- dh ih/
/dh uh/	/- dh uh/	/d ih/	/dh iy n/
/eh/	/ux/		

Table 6-2: 50 different ways *the* was pronounced, according to spectrogram readers.

30% if weighed by frequency, yet they are accountable for almost 50% of the errors.

In view of the above analysis, we propose a new speech unit: *function-word-dependent phone modeling*. Function-word-dependent phones are the same as word-dependent phones, except they are used only for function words. This improves the modeling of the most difficult subset of words, where the phones are most distorted. Unlike word-dependent phones, function-word-dependent phones are easily trainable because function words occur frequently in any task. For the same reason, they are also task-independent. Finally, for SPHINX, modeling function-word-dependent phones has the benefit of absorbing multiple pronunciations, which are not explicitly modeled.

We selected a set of 42 function words, for which we felt there was significant word-dependent co-articulatory effects, as well as adequate training data. A few of these words are not usually considered function words, but are appropriate for this task. These function words are shown in Table 6-3.

The training for function-word-dependent phone models was described in the previous section. The λs for the function-word-dependent models were distribution-dependent, because there was sufficient data for most distributions to estimate them. Table 6-4 shows the phones in *are* and *be*, the counts for the distributions of each phone, and the λs for the function-word-

A	ALL	AND	ANY	ARE	AT
BE	BEEN	BY	DID	FIND	FOR
FROM	GET	GIVE	HAS	HAVE	HOW
IN	IS	IT	LIST	MANY	MORE
OF	ON	ONE	OR	SHOW	THAN
THAT	THE	THEIR	TO	USE	WAS
WERE	WHAT	WHY	WILL	WITH	WOULD

Table 6-3: The list of 42 function words that SPHINX models separately.

dependent model parameters, phone model parameters, and uniform distribution are also shown. It can be seen that some distributions, such as the distributions between /b/ and /iy/ in *be*, are much more dependent on word context than their counts would have indicated.

Word	Phone	Dist.	Count	λ_{wdep}	λ_{indep}	$\lambda_{uniform}$
ARE	/aa/	Begin	2333	0.788	0.173	0.039
		Middle	2025	0.706	0.284	0.010
		End	1266	0.830	0.127	0.043
	/r/	Begin	1513	0.890	0.084	0.026
		Middle	1794	0.904	0.092	0.004
		End	2016	0.814	0.173	0.013
BE	/b/	Begin	176	0.207	0.786	0.007
		Middle	249	0.263	0.732	0.005
		End	243	0.705	0.295	0.000
	/iy/	Begin	222	0.636	0.358	0.006
		Middle	571	0.348	0.651	0.000
		End	329	0.337	0.659	0.004

Table 6-4: λs trained for phones in *be*. λ_{wdep} is the weight for function-word-dependent model, λ_{indep} is the weight for context-independent model, and $\lambda_{uniform}$ is the weight for uniform distribution.

6.4. Generalized Triphones

In our criticism of triphones, we argued that some phones have the same effect on neighboring phones. For example, the place of articulation has an important effect on the neighboring vowels. /b/ and /f/ have similar effects on the right-neighboring vowel, while /r/ and /w/ have similar effects on the right-neighboring vowel. Figure 6-3 illustrates this phenomenon. If we could identify these similar contexts, and merge them, we would have a much more manageable number of models, as well as much more training for each model.

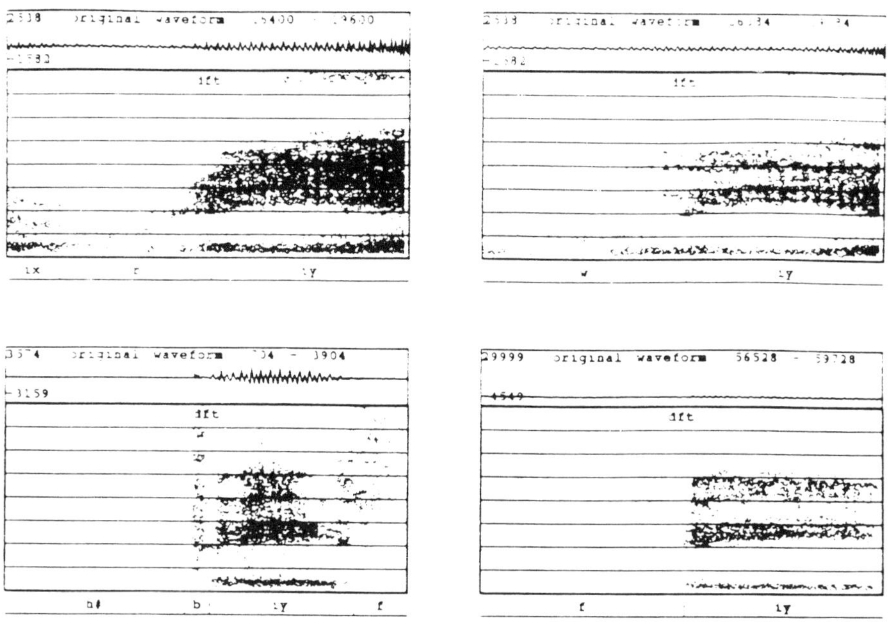

Figure 6-3: The waveforms and spectrograms for the phoneme /iy/ with four different *left-contexts* are illustrated. Note that /r/ and /w/ have similar effects on /iy/, while /b/ and /f/ have similar effects on /iy/. This illustrates that different left-contexts may have similar effects on a phone.

One approach is to merge perceptually similar contexts together using human knowledge [Derouault 87, Deng 88]. This guarantees that the merged contexts are sensible ones; however, if we were to consider all triphones, this would be a complicated process. Moreover, while the example we gave is clear, there are many where there may be no consensus even among the experts. Therefore, we believe it is desirable to automate this process of context generalization.

A greedy context merging procedure could be used to find and combine similar contexts:

1. An HMM is generated for every triphone context.
2. Clusters of triphones are created; initially, each clusters consists of one triphone.
3. Find the *most similar* pair of clusters which represent the same phone, and merge them together.
4. For each pair of clusters, consider moving every element from one to the other.
    1. Move the element if the resulting configuration is an improvement.
    2. Repeat until no such moves are left.
5. Until some convergence criterion is met, go to step 2.

Without step 4, this is simply an agglomerate clustering procedure [Duda 73], where every merge cannot be undone. Step 4 is a heuristic optimization that attempts to improve the clustering by allowing elements to be moved from one cluster to another. Although it appears expensive, if we remember which clusters have changed, and which cluster-pairs need not be compared, step 4 only triples the total computation for our task.

Many criteria could be used to determine the similarity between two HMMs. Juang and Rabiner [Juang 85c] proposed several similarity measures using cross entropy, divergence, and discrimination information. Paul and Martin [Paul 88] investigated merging of continuous HMMs using a chi-square measure. Finally, D'Orta *et al.* [D'Orta 87] proposed several measures, including output string/symbol probability, and maximum mutual information.

In this study, we use an information theoretic measure that determines the similarity between two HMMs based on the amount of information lost when the two models are merged. We use entropy of the original and merged HMMs to measure the information lost, and forward-backward counts to weigh the information lost. Entropy clustering has been used by Lucassen [Lucassen 83] to derive word baseforms from spelling, and by Brown [Brown 87] to merge similar VQ codewords.

While it is possible to compute the entropy of a Markov process [Hamming 86], it is a rather expensive process. Therefore, for the purpose of context clustering, we will ignore transition probabilities, and define the entropy of an HMM as the bits of information in the output pdf's. Let:

$N_{a,d}(i)$ be the count for codeword i in distribution d of context a of a phone as determined by the forward-backward algorithm.

$$N_{a,d} = \sum_i N_{a,d}(i) \qquad (3)$$

To normalize these counts into output probabilities:

$$P_{a,d}(i) = \frac{N_{a,d}(i)}{N_{a,d}} \qquad (4)$$

The entropy of an output pdf for distribution d for some phone in context a is defined as:

$$H_{a,d} = -\sum_i P_{a,d}(i) \cdot \log(P_{a,d}(i)) \qquad (5)$$

If we want to merge distribution d of two models that represent the same phone in context a and b into a merged model in context m, the new counts for distribution d in context m are simply:

$$N_{m,d}(i) = N_{a,d}(i) + N_{b,d}(i) \qquad (6)$$

We can compute $H_{b,d}$ and $H_{m,d}$ as we computed $H_{a,d}$. The information lost when a distribution d for context a and b are merged into m, weighted by counts, is:

$$L_d(a,b) = N_{m,d} H_{m,d} - N_{a,d} H_{a,d} - N_{b,d} H_{b,d} \qquad (7)$$

Finally, the information lost when two HMMs for context a and b are merged, weighted by counts, is:

$$L(a,b) = \sum_{d'} L_{d'}(a,b) \qquad (8)$$

Equation 8 is the distance metric used in our triphone clustering algorithm. This distance metric weighs the difference between models according to the frequency of the models. By preferring to merge models that do not appear frequently, each generalized model will be more trainable.

This context generalization algorithm provides the ideal means for finding the equilibrium between trainability and sensitivity. Given a fixed amount of training data, it is possible to find the largest number of trainable context models, or the smallest number of sensitive models. Armed with this technique, we could attack any problem and find the "right" number of models that are as sensitive and trainable as possible.

Table 6-5 gives the number of models created for each phone. It can be seen that the number of models of a phone is highly correlated with the frequency of the phone.

Table 6-6 shows the 19 clusters created for the phone /ae/ when the clustering process has reduced 2381 triphones to 500 triphone clusters. Most clusters consist of triphones that are easily identified as similar contexts.

6.5. Summary of SPHINX Training Procedure

Since our description of the training procedure in SPHINX has been fragmented, we now review the procedure. The entire training procedure is illustrated in Figure 6-4.

We first create 48 initial context-independent phone models from labeled TIMIT sentences. These phone models are then smoothed (to avoid mimicking pure phones) and used to initialize training on 4200 resource management sentences. The 48 trained context-independent phone models are then used to initialize the context-dependent phone models (which may be function-word-dependent phone models, generalized triphone models, etc.) for context-dependent training. A different pronunciation dictionary that uses the detailed models is used for the final forward-backward training. The final iteration of the forward-backward algorithm produces context-independent and context-dependent estimates, which are interploated with a uniform distribution. This set of interpolated context-dependent models is used for recognition.

Phone	Models	Phone	Models	Phone	Models
ae	37	l	47	p	21
eh	50	r	44	t	27
ih	34	w	14	f	21
iy	40	y	9	s	55
en	3	uh	3	sh	10
ah	18	er	38	th	12
ax	43	m	28	v	15
ix	30	n	44	z	30
aa	28	ng	5	hh	9
ao	14	ch	8	sil	12
uw	21	jh	10	dd	4
aw	7	b	17	pd	16
ay	19	d	20	td	8
ey	34	dh	3	kd	12
ow	24	g	10	dx	11
oy	1	k	34	ts	33

Table 6-5: Number of generalized triphone models for each phone when the total number of triphone models is 1000.

6.6. Results and Discussion

First, we present some results using previously proposed units of speech. These results are shown in Table 6-7. Left-context modeling of a phone considers two phones with the same identity to be different if their left-contexts are different. The left context of a phone is its left neighbor in the expected pronunciation, or a special symbol (#) when the phone is at the beginning of a word. Left-context dependent phone parameters are interpolated with context-independent phone parameters and a uniform distribution. Right-context modeling is the same, except right contexts are used. We see that the recognition rate improved substantially with the addition of left or right context, at the expense of 16 times as many models.

```
[ 1] (dh,td)
[ 2] (hh,v) (hh,td)
[ 3] (hh,z) (p,s)
[ 4] (k,l) (g,l) (#,p)
[ 5] (k,sh) (k,dx)
[ 6] (l,n) (l,m)
[ 7] (l,ch) (l,sh) (l,ts) (l,td) (l,dx) (l,s)
[ 8] (m,k) (m,kd) (s,k)
[ 9] (p,kd) (p,k) (t,k)
[10] (k,n) (k,m) (k,r) (z,m) (ch,n) (s,m)
[11] (r,th) (r,s) (r,f) (r,kd) (r,k)
[12] (#,dx) (#,dd) (hh,dd) (b,jh) (b,dd)
[13] (#,l) (#,f) (#,b) (b,sh) (m,dx) (n,sh) (hh,f)
[14] (#,ae) (ay,m) (jh,n) (iy,kd) (y,ng) (d,g)
 (jh,k) (r,dd)
[15] (#,td) (#,s)
[16] (#,v) (#,r)
[17] (v,l)
[18] (s,dx) (d,dx) (f,s) (b,s) (k,td) (t,s) (s,s)
 (ch,dx) (k,z) (sh,s) (g,s) (k,s) (k,t) (k,th)
[19] (z,n) (s,n) (v,n) (#,ng) (b,ng) (m,n) (f,n)
 (p,n) (r,n) (r,m) (er,m)
```

Table 6-6: 19 clusters created for the for (left, right) contexts for phone /ae/; # represents word-boundary context.

Triphone context considers two phones with the same identity to be different when either the left or the right context is different. Similar to BBN's approach [Schwartz 85], triphone models are interpolated with left-context phone models, right-context phone models, context-independent models, and a uniform distribution. The interpolation weights, however, were trained from deleted interpolation. Triphone models led to significantly better results than left or right context models.

Next, we implemented function-word-dependent phone modeling, and used it in conjunction with each of the versions above. The results with and without function-word-dependent phone modeling are shown in Table 6-8. We see that in each of the four cases, modeling function-word-dependent phones led to improvements. The improvement is the smallest for triphone contexts, because about half of the phones in function words have unique triphone contexts, which means triphone modeling was already doing

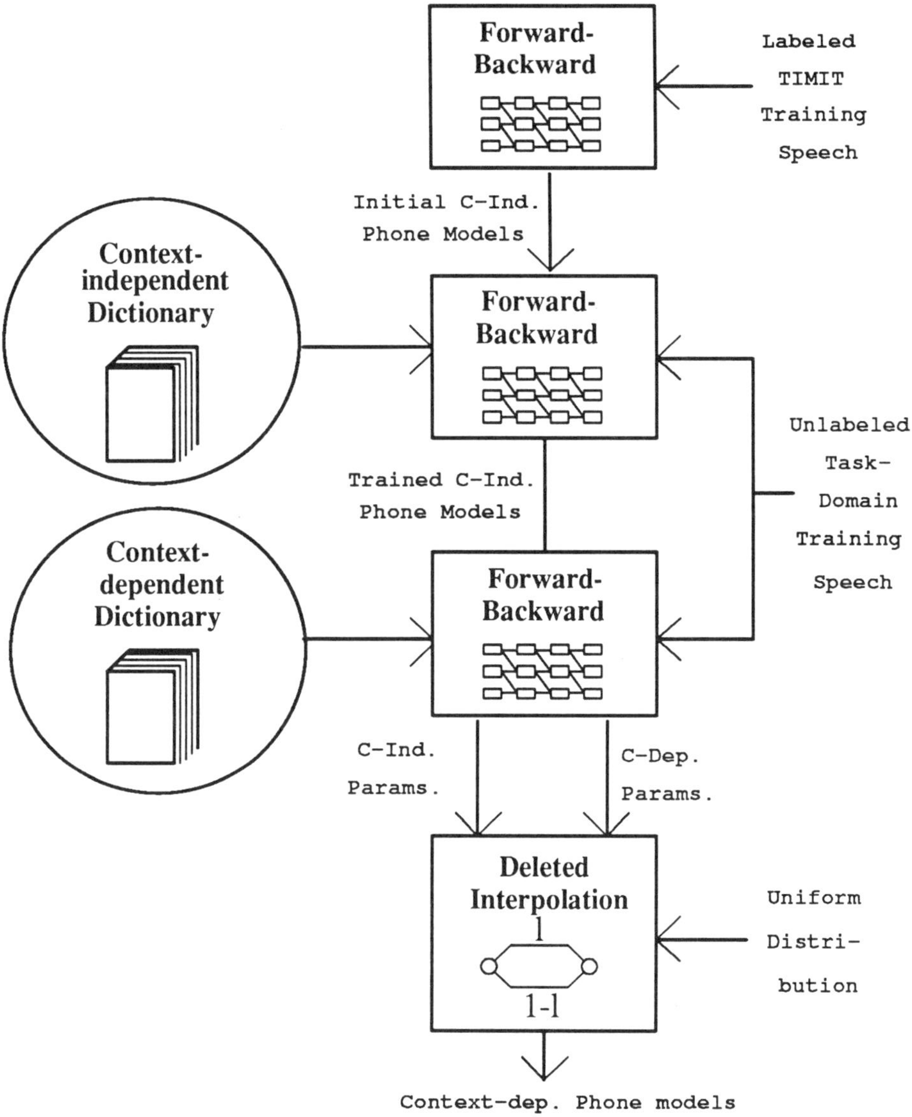

Figure 6-4: The training procedure in SPHINX.

function-word-dependent phone modeling for some function words. Nevertheless, these results clearly justify the utility of function-word-dependent phone modeling. We expect that function-word-dependent phone models could significantly improve systems that do not have very detailed phone models, or systems with very large vocabularies, where phones in function words would not be uniquely specified by triphones.

Version	Models	No grammar	Word pair	Bigram
Context-indep.	48	55.1% (49.6%)	86.8% (84.4%)	91.2% (90.6%)
Left-context	787	64.8% (61.6%)	90.5% (89.0%)	94.2% (93.8%)
Right-context	786	65.4% (62.1%)	91.0% (89.3%)	94.3% (94.0%)
Triphone	2381	72.7% (69.9%)	93.5% (92.2%)	95.4% (95.1%)

Table 6-7: Results with context-dependent and word-dependent phone modeling. Results shown are percent-correct (word-accuracy).

Version	Models	No grammar	Word pair	Bigram
Context-ind.	48	55.1% (49.6%)	86.8% (84.4%)	91.2% (90.6%)
CI+fnwd-dep.	153	62.9% (57.0%)	90.6% (87.9%)	93.8% (93.0%)
Left-context	787	64.8% (61.6%)	90.5% (89.0%)	94.2% (93.8%)
LC+fnwd-dep.	892	69.7% (66.6%)	92.7% (91.1%)	95.0% (94.7%)
Right-context	786	65.4% (62.1%)	91.0% (89.3%)	94.3% (94.0%)
RC+fnwd-dep.	891	70.2% (67.2%)	93.0% (91.5%)	95.1% (94.7%)
Triphone	2381	72.7% (69.9%)	93.5% (92.2%)	95.4% (95.1%)
TC+fnwd-dep	2447	72.8% (69.9%)	93.6% (92.4%)	95.5% (95.2%)

Table 6-8: Improvement from function-word-dependent phone modeling. Results shown are percent-correct (word-accuracy).

Table 6-9 gives the number of errors (substitutions + deletions + insertions) made by SPHINX (context-independent models, no grammar) with and without the use of function-word-dependent phone models. With function-word-dependent phone modeling, function word errors are cut by 27%, which accounts for almost all of the improvement from 45.3% to 53.4% accuracy.

Results for generalized triphone modeling are shown in Table 6-10. We ran the information-theoretic agglomerate clustering algorithm from 2381 triphones, and saved the clusters for every 100 merges. We then trained and tested on 10 new dictionaries, with 100, 200, 300, 400, 500, 600, 800, 1000, 1200, and 1400 models. We also include results with 48 HMMs (phone models, complete generalization), and 2381 HMMs (triphone models, no

Model Type	Function Word Errors	Other Errors
Context-ind.	357	350
CI+fnwd-dep.	261	334

Table 6-9: Number of function word errors and non-function-word errors with and without function-word-dependent phone modeling. Context-independent models were used without grammar.

generalization).

We see that in each case, the performance is superior to earlier results with comparable number of models. In particular, with only 100-200 models generalized triphones performed as well as 700-800 left or right context dependent phones. This clearly demonstrates the importance of modeling the left *and* right contexts, and the appropriateness of entropy clustering.

It is also interesting to note that the performance reaches an asymptote at about 1000 models. At 1000 models, an equilibrium between trainability and sensitivity was reached, given the amount of training. We believe the performance can be improved with more models and more training.

Finally, we added function-word-dependent phone modeling to triphone clustering, and ran an experiment with 1000 generalized triphone models plus 153 function-word-dependent phone models. Since some of the function-word-dependent phones are uniquely specified by the generalized triphones, there are a total of 1076 models. The results with these 1076 models are shown in Table 6-11.

6.7. Summary

In Chapter 1, we enumerated three reasons why continuous speech recognition is difficult. In this chapter, we introduced two classes of speech units to address these problems.

In order to deal with co-articulation, we used context-dependent triphone models. Since there are a large number of triphones, triphone models are poorly trained, and do not consider similar contexts. We used a

Number of gen. models	No grammar	Word pair	Bigram
48	55.1% (49.6%)	86.8% (84.4%)	91.2% (90.6%)
100	65.5% (60.9%)	90.8% (89.1%)	94.8% (94.0%)
200	69.0% (66.4%)	92.5% (91.0%)	94.9% (94.2%)
300	69.5% (66.2%)	92.5% (91.1%)	95.0% (94.1%)
400	70.5% (67.9%)	92.9% (91.8%)	95.0% (94.3%)
500	72.6% (69.6%)	93.0% (92.0%)	95.5% (95.1%)
600	73.3% (70.0%)	93.6% (92.4%)	95.5% (95.1%)
800	73.4% (70.3%)	93.7% (92.9%)	95.6% (95.1%)
1000	73.6% (70.3%)	94.2% (93.3%)	95.8% (95.4%)
1200	73.4% (70.0%)	94.0% (93.0%)	95.7% (95.3%)
1400	73.3% (69.7%)	93.8% (92.7%)	95.6% (95.2%)
2381	72.7% (69.9%)	93.5% (92.2%)	95.4% (95.1%)

Table 6-10: Results of generalized triphone modeling without function word modeling. Results shown are percent-correct (word-accuracy).

Version	Models	No grammar	Word pair	Bigram
Gen. Triphones	1000	73.6% (70.3%)	94.2% (93.3%)	95.8% (95.4%)
Gen. Triphones +fnwd-dep.	1076	74.2% (70.6%)	94.7% (93.7%)	96.2% (95.8%)

Table 6-11: Results with triphone context modeling. Results shown are percent-correct (word-accuracy).

procedure for generalizing similar contexts. Similar contexts for the same phone were identified and merged using an information theoretic criterion. This technique enabled us to automatically find the equilibrium between model trainability and model sensitivity for any given training database. Experiments showed that this led to large improvements over context-independent models, and significant improvements over other types of context-dependent models.

In order to deal with unclear function words in continuous speech, we modeled the phones in function words explicitly. We believe that function words are a class of words which have the greatest word-dependent co-articulatory effects. Other word-dependent phones can be approximated by slightly less appropriate, but better trained, context-dependent phone models. By modeling the phones in function words, we focus on the most difficult subset of any large vocabulary, as well as the subset in which the phones are most distorted. Compared to word-dependent phone modeling, function-word-dependent phone modeling has additional benefits of task-independence and trainability because of their high frequency of occurrence. We showed that using function-word-dependent models alone improves the recognition rate significantly, although less than triphone modeling. By combining 1000 generalized triphone models and function-word-dependent phone models, some additional improvement was obtained; however, this improvement was considerably smaller than that obtained from adding function-word-dependent phone models to less detailed context models. This is because, for our task, many function-word-dependent phones were uniquely specified by their corresponding generalized triphones. For a larger task, we expect that this will no longer be the case, and that function-word-dependent phone modeling will complement triphone models.

The final problem of unclear word boundaries is not addressed explicitly in this study. By training without accounting for word-boundaries, juncture effects were implicitly absorbed in our models. We could easily incorporate juncture rules in our training procedure by allowing substitution or deletion of certain phones near particular word boundaries. However, it is not so simple to incorporate juncture rules in a recognizer because the number of juncture phenomena could be as large as the square of the number of words when no grammar is used. The speed of the recognizer would also be substantially slower. Therefore, for practical reasons, we chose not to address this issue, although it will be investigated when we extend this work.

Instead of using hand-tuned weights to average the different models together, we use deleted interpolated estimation to automatically assign weights to the different estimates based on how well each estimate predicts unseen data.

In summary, we examined many of the currently used speech units, and assessed the benefits and shortcomings of these units. We proposed two new units, function-word-dependent phone modeling and generalized triphone modeling. We also applied the deleted interpolated estimation algorithm to combine models that have differing degrees of robustness and appropriateness. We showed that these techniques result in very substantial improvements over context-independent models.

7 Learning and Adaptation

There are many anatomical differences between speakers, such as the length of the vocal tract, the size of the nasal cavity, etc.. There are also many speaker-dependent speaking habits, such as accent, speed, loudness, etc.. These differences have been considered so extreme that many researchers have forsaken speaker-independent recognition. In the preceding chapters, we have shown that accurate speaker-independent recognition is possible, and that one need not be so pessimistic. However, we do acknowledge these speaker differences, and certainly believe that blindly mixing the parameters of all the speakers discards useful information.

In a speaker-independent recognizer, the use of speech from many speakers enables reliable and robust estimation of a large number of parameters. However, when many distributions are broad, it is difficult to make fine distinctions. And when the distributions are multi-modal, nonsense matches will be considered, and possibly recognized. For example, the first half of a word may be matched against the female half of a bimodal distribution, while the second half of the word may be matched against the male half of the distribution.

In view of these problems, it would be desirable to normalize for or adapt to these speaker characteristics. Ideally, we would like to retain the robustness of well-trained models, yet improve the appropriateness of these models given some knowledge of a speaker. This knowledge could only come from adaptation sentences produced by the speaker. In this chapter, we present two adaptation algorithms.

7.1. Speaker Adaptation through Speaker Cluster Selection

The first adaptation algorithm is predicated upon the assumption that it is possible to divide the class of all speakers into *speaker clusters*, so that each speaker cluster contains similar speakers.

This assumption suggests that we could divide the training speakers into clusters, and train a set of VQ codebooks and hidden Markov models for each cluster. Prior to recognition, a *cluster identification* stage is needed. During this stage, the input speaker is identified as a member of the cluster that best resembles him/her. Then, the HMMs for that cluster are instantiated into the recognizer, and recognition proceeds as in the speaker-independent mode. This is illustrated in Figure 7-1. This scheme requires two major modifications of standard training and recognition strategy: (1) how to divide the training speakers into clusters, and (2) how to identify the best speaker cluster for the input speaker.

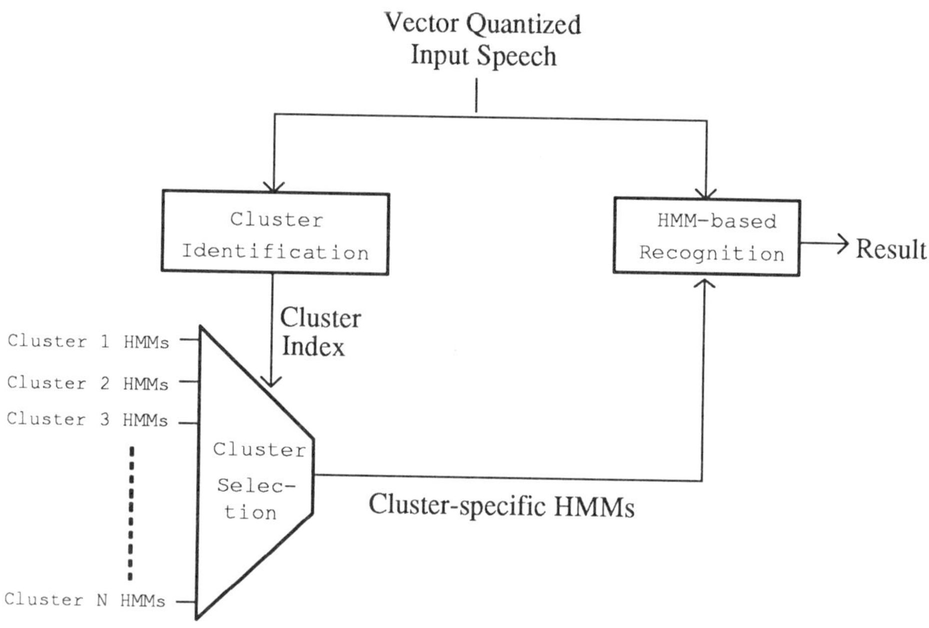

Figure 7-1: Speaker adaptation by cluster selection.

7.1.1. Speaker Clustering

Prior to adaptation, we first divided all training speakers into speaker clusters. Since we only have 105 speakers, it is possible to use the modified agglomerate clustering that we used for context clustering. Each of the 105 speakers was originally placed in his/her own cluster. For each speaker, we generated 48 phonetic HMM from his/her 40 sentences of training speech. To save computation, only 21 of these HMMs were used for speaker clustering. Each HMM had nine distributions, leading to 189 discrete distributions per speaker.

The clustering algorithm was then run to iteratively merge the two speaker clusters that resulted in the least loss of information, and then move elements from cluster to cluster to improve the total information. This process continued until two clusters were left. This left us with as few clusters as two, and as many as 104.

Table 7-1 shows the male/female composition of the clusters created by this algorithm. Most clusters are dominated by male or female. When two clusters were left, one cluster was completely male, and the other completely female. This is a very positive sign for our clustering algorithm and metric.

	Number of Clusters Left											
Cluster	7		6		5		4		3		2	
Number	M	F	M	F	M	F	M	F	M	F	M	F
1	10	2	10	2	10	2	10	2	0	28	0	30
2	6	1	6	1	0	27	0	27	33	1	75	0
3	0	7	0	27	27	0	23	0	42	1		
4	0	20	26	0	21	0	42	1				
5	26	0	20	0	17	1						
6	20	0	13	0								
7	13	0										

Table 7-1: Male/female composition of the 2 to 10 clusters as produced by the agglomerate clustering algorithm.

Ideally, we would like to have a large number of clusters. However, with only 105 training speakers, if we generate too many clusters, we would have estimation problems with clusters that contained too few speakers. Thus, to ensure that our VQ codebooks and HMMs are well-trained, we chose to use three speaker clusters. The first cluster consisted of 28 female speakers. The second cluster contained 1 female and 33 male speakers, and the last cluster contained 1 female and 42 male speakers. For each of the three clusters, we generated cluster-dependent VQ codebooks and HMMs.

7.1.2. Speaker Cluster Identification

Before cluster-dependent recognition could commence, the input speaker was first classified into one of the speaker clusters. Our classification scheme is very simple: the input speaker must produce a training sentence, as well as the identity of the sentence. This sentence was vector quantized into each of the three cluster-dependent codebooks, and force-aligned against the expected sequence of words. The forced-alignment was implemented as an iteration of the forward algorithm. The speaker cluster that produced the largest probability (alpha-terminal) from the forced-alignment was identified as the input speaker's best matching cluster. Subsequently, cluster-dependent codebooks and parameters were used for the input speaker.

Note that although supervised training is needed for one sentence, it is not necessarily burdensome, because SPHINX could initially operate in a speaker-independent mode, and use any sentence for cluster selection. For almost all tasks, it is possible to know or infer that a sentence has been correctly recognized. For many tasks, the true identity of that sentence must be provided when misrecognition occurs. Therefore, it is possible to either wait until a sentence has been correctly recognized, or use immediate user feedback to begin the adaptation process.

7.2. Interpolated Re-estimation of HMM Parameters

The cluster selection algorithm introduced in the previous section uses *similar speaker clusters* to adapt SPHINX to an input speaker. If unlimited training data were available, a large number of well-trained clusters could be generated. In that case, *speaker cluster selection* algorithms should do an adequate job of adaptation. However, we only have enough training data to create three clusters. While these three clusters are better representations than speaker-independent training, they are still far from speaker-dependent training. Therefore, it would be desirable to adapt to an individual speaker, given a small number of adaptive-training sentences from that speaker.[16]

With only a few adaptation sentences, it would not be possible to create a speaker-dependent codebook. Therefore, we will assume a fixed existing codebook, which could be speaker-independent, or cluster-dependent. Adaptation for an individual speaker will then involve improving the parameters given this fixed codebook.

The obvious way to obtain parameters most suitable for the individual speaker is to train on the adaptation sentences. But these parameters will be very poorly estimated because of limited training. Since we have well-trained speaker-independent parameters, we could combine the two into estimates that are more suitable than speaker-independent parameters, yet more robust than speaker-dependent ones. Once again, we find an ideal application for *deleted interpolation*.

But we can do better than just combining these two measures. From the adaptation sentences, we can train other estimates with various degrees of reliability and robustness. In our current implementation, there are five measures: (1) speaker-dependent parameters, (2) speaker-independent parameters, (3) speaker-dependent tied parameters, (4) co-occurrence smoothed parameters, and (5) similar speaker parameters.

7.2.1. Different Speaker-Adaptive Estimates

Speaker-Independent and Speaker-Dependent Parameters

Speaker-independent parameters are directly taken from the speaker-independent SPHINX System, described in earlier chapters. Speaker-dependent parameters are derived by running the forward-backward algorithm on the adaptation sentences.

Speaker-Dependent Tied Parameters

In our current phone models, each phone is assumed to have three pdf's. By tying all three pdf's for each phone together, we would improve the robustness of our speaker-dependent estimates by sacrificing some accuracy. These tied parameters are another set of estimates.

---

[16]With a large amount of training, we would simply train speaker-dependent models.

Co-occurrence Smoothed Parameters

The speaker-dependent parameters suffer from sparseness. Suppose several codewords *should* be likely for a distribution. In reality, perhaps only one of them is observed, while the others are zeroes. If we have a similarity measure among all codewords, we could use it to replace these zeroes with appropriate larger numbers. This enables us to differentiate the *likely* unobserved codewords from the *unlikely* ones. For this purpose, we introduce the co-occurrence smoothing algorithm.

We define $CP(i \mid j)$, the *co-occurrence probability* of codeword i given codeword j, as[17]:

$$CP(i \mid j) = \frac{\sum_{p=1}^{NP} \sum_{d=1}^{ND(p)} P(i \mid p,d) \cdot P(j \mid p,d) \cdot P(p) \cdot P(d)}{\sum_{k=1}^{NC} \sum_{p=1}^{NP} \sum_{d=1}^{ND(p)} P(k \mid p,d) \cdot P(j \mid p,d) \cdot P(p) \cdot P(d)} \qquad (1)$$

where NP is the number of phonemes, $ND(p)$ is the number of output pdf's in the HMM for phoneme p, NC is the number of codewords in the codebook, and $P(k \mid p,d)$ is the output probability of codeword k for distribution d in phoneme model p. *Co-occurrence probability* can be loosely defined as "when codeword j is observed, how often is codeword i observed in similar contexts." In our definition, "similar context" means the same output pdf.

We could use the *co-occurrence probability* (CP) to smooth the speaker-dependent output pdf's (P) into a smoothed pdf (SP):

$$SP(k \mid p,d) = \sum_{i=1}^{NC} CP(k \mid i) \cdot P(i \mid p,d) \qquad (2)$$

SP achieves the desired effect of making speaker-dependent parameters more robust. The smoothing effect is illustrated in the two sets of pdf's in Figure 7-2. More details on co-occurrence smoothing can be found in [Lee 88b].

---

[17]The *co-occurrence probabilities* can be more conveniently computed from the counts accumulated in forward-backward by a simple transformation of the equation.

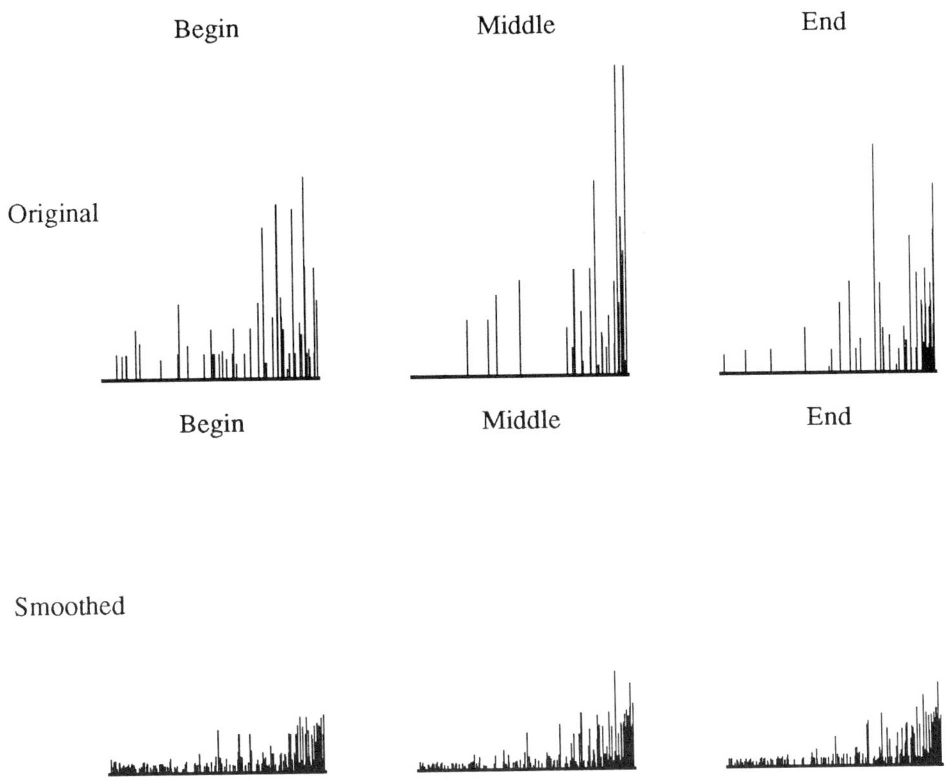

Figure 7-2: The effect of co-occurrence smoothing. The top pdf's represent the cepstrum codebook of a poorly trained model (*P*). The second set of pdf's has been smoothed (*SP*).

Similar-Speaker Parameters

The tied distribution and co-occurrence smoothing techniques derived more robust but less accurate estimates from speaker-*dependent* parameters. We can also derive more accurate but less robust estimates from speaker-*independent* parameters. The speaker-independent parameters were generated from many speakers, some of whom are similar to the input speaker, while others are very different. We could select a set of training speakers who are similar to the input speaker, and use these speakers to generate a set of *similar-speaker* estimates. Another advantage for using *similar-speaker* parameters is that even if a phone was never observed in the adaptation sentences, the combination with the phone parameters derived from *similar speakers* should improve the speaker-independent parameters.

To find these similar speakers, we first generated a set of phonetic HMMs from each training speaker's speech. These poorly trained HMMs were smoothed with the speaker-independent models using deleted interpolation. Then, we took a subset of the input speaker's adaptation sentences, and force-aligned them with each training speaker's HMMs using the forward algorithm. The 20 speakers whose HMMs yielded the highest probability from the forward algorithm comprise the set of similar speakers.

The training speech for these 20 speakers were used to train *similar-speaker parameters*. If time were a major problem, we could simply sum and then normalize the forward-backward counts for these 20 speakers. In this study, we actually re-ran two passes of the forward-backward algorithm on the 20 similar speakers to obtain the *similar-speaker parameters*.

7.2.2. Interpolated Re-estimation

To combine these five estimates, we first trained a set of λs using deleted interpolation. We divided the training speakers into two blocks. We trained a set of HMMs from the first block. For each speaker in the second block, we computed the five estimates from S sentences, and estimated λs using the remaining $40-S$ sentences (where S is the number of adaptation sentences to be used). A λ was computed for each estimate, and for each range of speaker-dependent counts. When the speaker-dependent count was high, the speaker-dependent parameters should dominate. In general, more appropriate estimates should have higher λs. Table 7-2 shows the weights determined for each range and estimate. Among the three indirectly derived estimates, tied distributions were not particularly good while co-occurrence smoothed estimates and similar speaker estimates were highly informative. The reason that tied distributions did not contribute much is probably due to our use of compound phones, whose three distributions are quite different, and should not be tied.

After the λs have been trained, for each input speaker, the five estimates were derived from his/her adaptation sentences, and then combined using these λs. Figure 7-3 illustrates the interpolated re-estimation learning in SPHINX.

Count Range	$\lambda_{Spk-dep.}$	λ_{Tied}	$\lambda_{Co-occur}$	$\lambda_{Spk-ind.}$	$\lambda_{Sim-spk.}$
$1 > Count \geq 0$	0.000	0.012	0.376	0.345	0.266
$10 > Count \geq 1$	0.119	0.010	0.201	0.428	0.242
$20 > Count \geq 10$	0.130	0.008	0.189	0.400	0.273
$30 > Count \geq 20$	0.149	0.009	0.088	0.324	0.441
$50 > Count \geq 30$	0.226	0.003	0.104	0.257	0.411
$80 > Count \geq 50$	0.251	0.004	0.106	0.096	0.543
$120 > Count \geq 80$	0.321	0.009	0.130	0.071	0.469
$\infty > Count \geq 120$	0.526	0.001	0.091	0.017	0.366

Table 7-2: The λs trained from deleted-interpolation for each set of parameters.

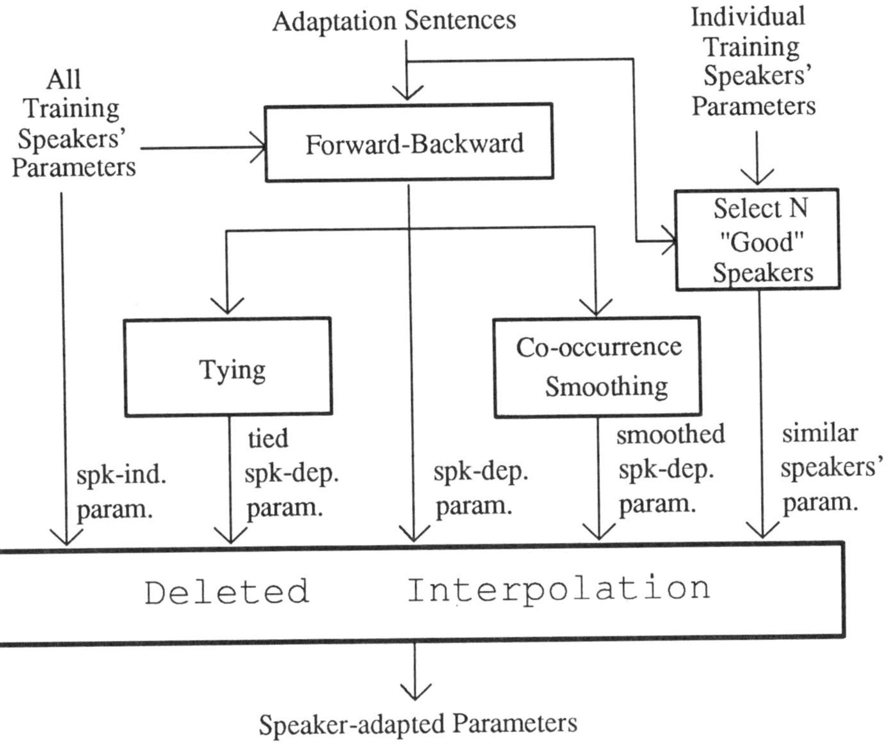

Figure 7-3: Interpolated Re-estimation Learning in SPHINX.

7.3. Results and Discussion

We implemented the speaker cluster selection algorithm with three speaker clusters. For each test speaker, we used one adaptation sentence to find the most appropriate speaker cluster for him/her. We then used the HMMs and codebooks for that cluster to recognize the test speaker's speech. We also tried to use HMMs and codebooks for the other two clusters. The word accuracies with context-independent phone models and with 600 generalized triphone models are shown in Table 7-3 and 7-4, respectively.

Codebook & HMM	No Grammar	Word Pair	Bigram Grammar
Best Cluster	49.5%	84.6%	90.6%
Second Cluster	41.8%	80.9%	87.1%
Worst Cluster	31.5%	70.3%	82.5%
Speaker-Indep.	49.6%	84.4%	90.6%

Table 7-3: Speaker cluster selection results using 48 phonetic models. Results shown are word accuracy.

Codebook & HMM	No Grammar	Word Pair	Bigram Grammar
Best Cluster	68.9%	92.1%	94.6%
Second Cluster	61.7%	88.2%	92.0%
Worst Cluster	52.5%	80.9%	88.3%
Speaker-Indep.	70.0%	92.4%	95.1%

Table 7-4: Speaker cluster selection results using 600 generalized triphone models. Results shown are word accuracy.

We believe that the cluster selection process functioned properly. In terms of VQ distortion, per-frame probability, and recognition rate, the best cluster was substantially better than the second best, which was substantially better than the worst cluster. However, the performance with the best cluster was not better than speaker-independent codebooks and HMMs, which mixed all the parameters together. In fact, with 600 generalized triphone models, speaker-independent codebooks and HMMs consistently outperformed the cluster-specific ones. This indicates that although the speaker-independent codebooks and HMM are not as appropriate as the cluster-dependent ones, the availability of more training data compensated for these deficiencies. The availability of more training was even more important when the number of models increased to 600. Although this result is disappointing, it is consistent with the fact that our speaker-independent system trained on 4200 sentences performs at the same level as BBN's speaker-dependent system trained on 600 sentences.

For interpolated re-estimation, we withheld 25 speakers from speaker-independent training, and estimated λs using deleted interpolation. Because of high computational costs, we did not actually delete different sets of 25 speakers.

For speaker-adaptive recognition using interpolated re-estimation, we computed speaker-adaptive estimates for each of the fifteen test speakers with 10 and 30 adaptation sentences, and combined them using the λ values learned from deleted interpolation. Table 7-5 gives the results of no adaptation, 10-sentence adaptation, and 30-sentence adaptation with context-independent phone models. There are appreciable improvements for each grammar from no adaptation to 10 adaptation sentences, and from 10 to 30. Table 7-6 gives the results with 1076 context-dependent phone models, which is the best SPHINX configuration described in the previous chapter. With 30 adaptation sentences, we were able to reduce the error rate by about 5–10%. When a grammar was used, the improvement was smaller. This is probably because robustness is more important with a grammar than without, and speaker adaptation does not improve the robustness of the speaker-independent models.

Condition	No grammar	Word Pair	Bigram
No adaptation	55.1% (49.6%)	86.8% (84.4%)	92.1% (91.5%)
10-sent. adapt.	59.3% (52.3%)	87.9% (85.6%)	92.5% (92.0%)
30-sent. adapt.	62.5% (57.5%)	89.2% (87.8%)	94.2% (93.0%)

Table 7-5: Speaker adaptation using interpolated re-estimation with 10 and 30 adaptation sentences.

Condition	No grammar	Word Pair	Bigram
No adaptation	74.2% (70.6%)	94.7% (93.7%)	96.2% (95.8%)
10-sent. adapt.	75.2% (71.3%)	94.6% (93.8%)	96.4% (95.8%)
30-sent. adapt.	76.5% (74.1%)	94.9% (94.0%)	96.9% (96.1%)

Table 7-6: Speaker adaptation using interpolated re-estimation with 10 and 30 adaptation sentences.

7.4. Summary

Most adaptation [Shikano 86a, Schwartz 87] algorithms adapt from parameters of a different speaker. They require a few adaptation sentences from the input speaker, as well as considerable time to perform the adaptation and mapping. Since the speaker-independent SPHINX System was already very accurate, our adaptation process could be a non-intrusive one in which we begin with a speaker-independent system, and gradually modify the parameters to better suit the input speaker.

In this chapter, we described two such algorithms. The first algorithm is speaker cluster selection adaptation. We divided the training speakers into three clusters, and trained cluster-dependent codebooks and HMMs. We were able to generate HMMs that were sharper and more appropriate for each cluster. However, because each cluster has only one-third as much training data as the speaker-independent codebooks and HMMs, the results were about the same as that obtained with speaker-independent codebooks and HMMs.

This disappointing result has not shattered our hopes for codebook-based adaptation. If we had a much larger database (of, say, 2000 speakers) which has probably well saturated the speaker-independent performance, then this clustering process will most likely pay off. Alternatively, we could try a variation of the codebook-normalization procedure [Shikano 86a, Schwartz 87, Feng 88]. With such a procedure, we would still generate speaker clusters and cluster-dependent codebooks. A probabilistic mapping from the speaker-independent codebook to the cluster-dependent codebook can be determined using the forward-backward algorithm. We could then use all of the training speakers to train a set of mappings for each cluster, thereby effectively using all the training speakers. That algorithm, however, relies on the accuracy of the mapping process.

The second algorithm is interpolated re-estimation. Given a small number of adaptation sentences, we combined speaker-independent parameters, speaker-dependent parameters, and other parameters derived from them using deleted interpolation. The speaker-adapted parameters were more well-trained than speaker-dependent parameters, yet more accurate than speaker-independent ones. We were able to reduce the error rate of the speaker-independent SPHINX System by about 5–10% using 30 adaptation sentences. We believe this algorithm has the potential of improving *speaker-dependent* systems by adding an interpolation stage after speaker-dependent training.

In this chapter, we introduced two algorithms that adapt at the codebook parameter level. The speaker cluster selection algorithm resulted in no improvement over speaker-independent recognition, while the interpolated re-estimation algorithm resulted in a significant but modest improvement. We did not get a larger improvement because our speaker-independent system is robustly trained, and because our adaptation algorithms require more training data than is currently available. In the future, we will explore an adaptive system that begins in a speaker-independent mode and adapts on a large amount of speaker-dependent training data. We will also pursue adaptation with limited enrollment at the speech parameter level [Brown 83, Stern 83], instead of at the codebook parameter level.

8 Summary of Results

8.1. SPHINX Results

Figure 8-1 shows improvements from earlier versions of SPHINX. The seven versions in Figure 8-1 correspond to the following descriptions with incremental improvements:

1. The baseline system, which uses only LPC cepstral parameters in one codebook.
2. The addition of differenced LPC cepstral coefficients, power, and differenced power in one codebook.
3. All four feature sets were used in three separate codebooks. This version was reported in [Lee 87], the first description of the SPHINX System.
4. Tuning of phone models and the pronunciation dictionary, and the use of word duration modeling.
5. Function word dependent phone modeling. This version was reported in [Lee 88a].
6. Generalized triphone modeling. This version was reported in [Lee 88c].
7. Interpolated re-estimation with 30 adaptation sentences.

Table 8-1 shows the recognition results for the 15 testing speakers. Although the performance appears variable from speaker to speaker, this variability is not predictable from the gender or the dialect of the speakers.

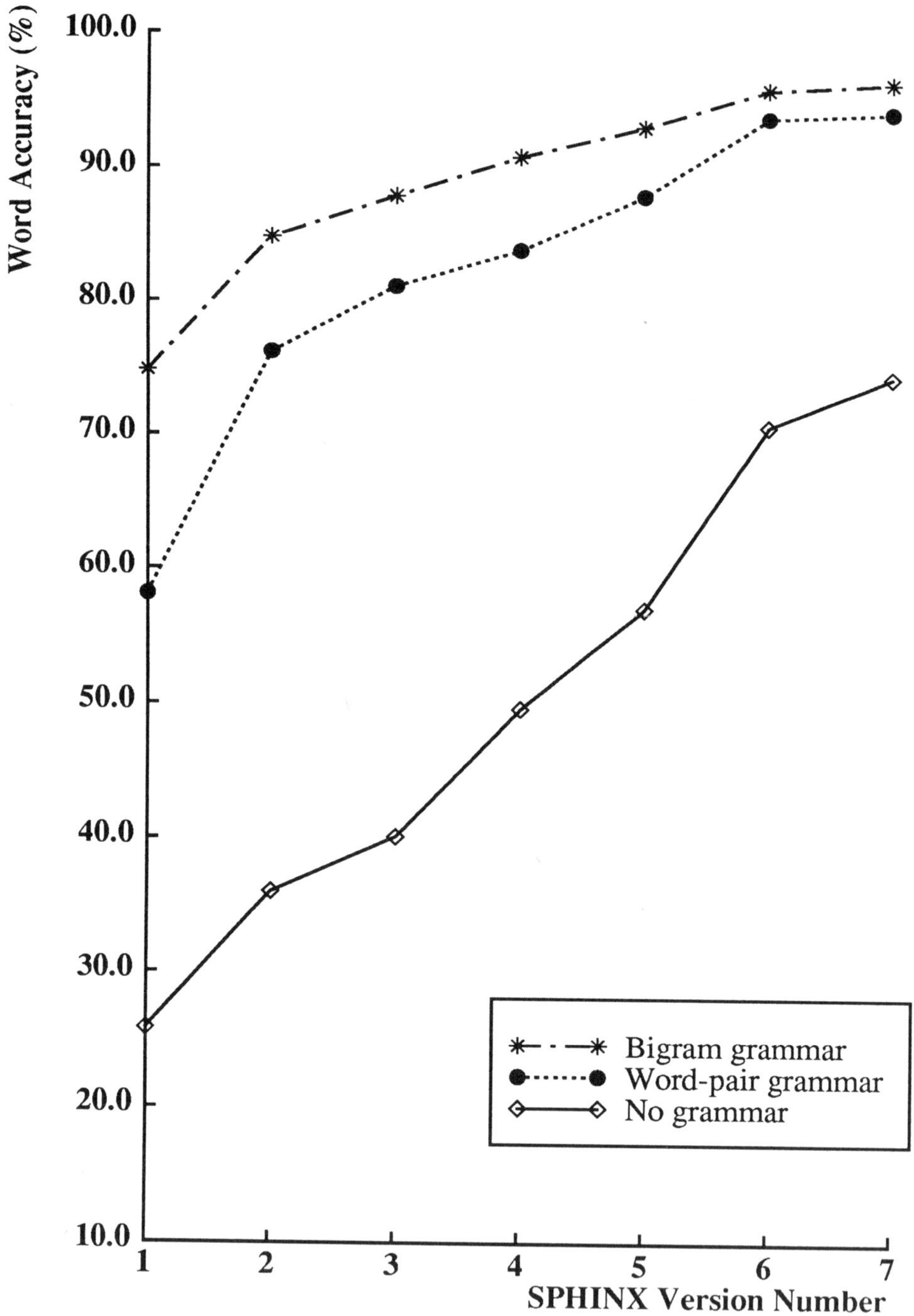

Figure 8-1: Results of five versions of SPHINX.

Initials	Gender	Dialect	No Gram.	W-Pair	Bigram
bcg	F	Moved	61.7%	91.9%	97.7%
sah	F	New Eng.	61.4%	91.0%	94.4%
ljd	F	North Mid.	67.3%	93.7%	98.2%
lmk	F	South	71.3%	96.6%	96.6%
awf	F	South	73.9%	94.4%	95.5%
dpk	M	New Eng.	65.5%	90.2%	93.9%
dab	M	New Eng.	71.3%	95.5%	98.5%
dlc	M	North Mid.	92.6%	100.0%	100.0%
gwt	M	Northern	83.2%	96.4%	97.6%
ctm	M	Northern	72.7%	89.3%	92.9%
jfc	M	NYC	61.2%	93.4%	88.9%
sjk	M	NYC	80.3%	95.1%	96.3%
ctt	M	South	73.6%	94.3%	98.9%
bth	M	Western	69.8%	97.7%	96.6%
jfr	M	Western	62.0%	88.1%	92.4%

Table 8-1: SPHINX word accuracy by speakers.

8.2. Comparison with Other Systems

In this section, we present results from other similar systems (about 1000-word vocabulary, continuous speech recognizers). Ideally, we would like to compare different systems based on the quality of their acoustic modeling alone; however, there are many other factors that contribute to the accuracy of a speech recognizer—language modeling, speech quality, speaker quality, vocabulary confusabiity, language used, etc.. Thus, these comparisons should not be construed as definitive statements about the superiority of some systems over others.

The two earliest systems that recognized continuous speech from a large vocabulary are HEARSAY [Lesser 75, Reddy 77] and HARPY [Lowerre 76, Reddy 77]. Both of these systems were speaker-*dependent*, and had percent correct rates of 86% and 97%, respectively. It should be noted that the document retrieval grammar used by HEARSAY and HARPY to achieve this recognition rate is very tight, with a perplexity of only 4.5. When extended to larger grammars, HARPY's performance degraded substantially [Reddy 77].

Before IBM began their colossal TANGORA project, they worked on the *Laser Patent* task, a 1000-word speaker-*dependent* continuous speech task [Bahl 80a]. The laser patent task had a perplexity of 24.5. The IBM-LASER system pioneered the use of vector quantization with discrete phonetic HMMs. It achieved a word accuracy of 88.6%.

More recently, European researchers at Siemens, Philips, and IPO collaborated in building SPICOS [Noll 87, Ney 88], a speaker-*dependent* continuous speech recognition system with a vocabulary of 917 German words, and a grammar with perplexity 74. Continuous Laplacian mixture densities were used with phoneme models. A word accuracy of 92% was reported.

BYBLOS [Chow 87, Kubala 88] is one of the most accurate speaker-*dependent* continuous speech recognizer today. Through extensive context-dependent modeling, it achieved accuracies of 92.5% and 67.6% for grammars with perplexity 60 and 997. Since these are exactly the same grammar and vocabulary that we used, a comparison between SPHINX and BYBLOS is more valid.

All of the aforementioned systems are speaker-*dependent*. Several other speaker-*independent* systems have emerged as part of the recent DARPA speech effort. In particular, results on the resource management task are available for CMU's ANGEL System [Adams 86] based on knowledge engineering techniques, a TI System [TI 87] based on continuous density word HMMs, and the SRI System [Murveit 88] based on discrete density phone HMMs. ANGEL's word accuracy is 41% with a grammar of perplexity 34. TI and SRI reported word accuracies of 44.3% and 40.4% with no grammar, respectively.

Table 8-2 summarizes these systems, their performance, as well as the performance of SPHINX. Table 8-2 shows that without adaptation, SPHINX is as already comparable to the best *speaker-dependent* systems.

System	Speaker	Voc. Size	Perplexity	% Corr.	Word Acc.
HEARSAY	dependent	1011	4.5	86%	-----
HARPY	dependent	1011	4.5	97%	-----
IBM-LASER	dependent	1000	24	91.1%	88.6%
BYBLOS	dependent	997	60	94.8%	92.5%
BYBLOS	dependent	997	997	70.1%	67.6%
SPICOS	dependent	917	74	92.8%	92.0%
ANGEL	independent	997	34	45.5%	41.0%
TI	indep-male	997	997	55.5%	44.3%
SRI	independent	997	997	43.6%	40.4%
SPHINX	independent	997	20	96.2%	95.8%
SPHINX	independent	997	60	94.7%	93.7%
SPHINX	independent	997	997	73.6%	70.6%

Table 8-2: Results of SPHINX compared to other similar systems.

8.3. Error Analysis

Appendix III contains the sentence-by-sentence recognition results of the best configuration of the speaker-independent SPHINX. In this section, we will look at the types of errors made by SPHINX. These analyses were performed after the results from the system have finalized.

We computed the number of times each word was inserted, deleted, and substituted for all three grammars. The words with error counts greater than one are shown in Table 8-3, 8-4, and 8-5.

In general, we found that most of the errors are reasonable confusions between similar words or sequences of words, such as *arriving* → *arrive in*, *were in* → *weren't*, *on first* → *Connifer's*, *that are in* → *centering*. These types of errors are most frequent when no language model was used, because there were many more combinations of word sequences that may be confusable. When the language model match factor (see Section 4.6) was large, these errors tended to merge smaller words (or phones) into larger ones; when the language model match factor was small, these errors tended to split larger words (or phones) into smaller ones. We have found that better

Bigram Grammar		
Insertions	**Deletions**	**Substitutions**
THE (2)	THE (13) ARE (2)	THERE (2) ON (2) OF (2) A (2) AND (2) ARE (2)

Table 8-3: The number of times words are inserted, deleted, and misrecognized in SPHINX using the bigram grammar. Only words with two or more errors are shown.

Word Pair Grammar		
Insertions	**Deletions**	**Substitutions**
	THE (17) A (3) IS (2)	THE (5) ON (4) ARE (3) AND (3) THAT (3) THERE (2) OF (2) A (2)

Table 8-4: The number of times words are inserted, deleted, and misrecognized in SPHINX using the word-pair grammar. Only words with two or more errors are shown.

recognition was obtained with a larger language model match factor, which prefers longer words over shorter ones, and deletions over insertions.

We also noted that although our explicit modeling of phones in function words has reduced function word errors substantially, these words were still the primary source of problems. When a grammar was used, most errors were misrecognitions or deletions of function words. When no grammar was used, more other types of errors occurred, but most errors were still misrecognitions or deletions of function words. The recognition of function words was particularly difficult with sequences of function words, such as *that are in the* or *less than that of the*.

No Grammar		
Insertions	**Deletions**	**Substitutions**
OF (4)	THE (22)	IN (24)
THE (3)	IN (10)	ARE (13)
TWO (3)	OF (9)	OF (13)
IF (2)	THAT (8)	THE (12)
IN (2)	A (7)	AND (11)
OFF (2)	ARE (5)	ON (10)
	AND (4)	IS (6)
	WAS (2)	AT (4)
	IS (2)	THAT (4)
	TO (2)	WHAT (4)
		FOR (4)
		GET (4)
		DATA (3)
		ANY (3)
		SOON (2)
		THERE (2)
		PERCENT (2)
		COPELAND (2)
		EIGHTY (2)
		THAN (2)
		ARRIVE (2)
		SAME (2)
		DO (2)
		AN (2)
		CURRENT (2)
		SET (2)
		EIGHT (2)
		OFF (2)
		HUNDRED (2)
		ZULU (2)

Table 8-5: The number of times words are inserted, deleted, and misrecognized in SPHINX using no grammar. Only words with two or more errors are shown.

A small portion of the errors could be attributed to the lack of juncture modeling. We have found that geminations did not cause as much problem as deletions. Finally, some errors might be corrected by providing alternate pronunciations. But as we found earlier, alternate pronunciations will correct some problems, but also introduce new ones because of the increased vocabulary size.

9 Conclusion

9.1. Trainability vs. Specificity : A Unified View

This monograph addressed the problem of large-vocabulary speaker-independent continuous speech recognition. At the outset, we chose to use hidden Markov modeling, a powerful mathematical learning paradigm. We also decided to use vector quantization and discrete HMMs for expedience and practicality. Then we attacked the problems of large vocabulary, speaker independence, and continuous speech within our discrete HMM framework.

It is well known that HMMs will perform better with detailed models. It is also well known that HMMs need considerable training. This need is accentuated in large-vocabulary, speaker-independence, and discrete HMMs. However, given a fixed amount of training data, model specificity and model trainability are two incompatible goals. More specificity usually reduces trainability, and increased trainability usually results in over-generality.

Thus, our work can be viewed as finding an equilibrium between specificity and trainability. To improve trainability, we used one of the largest speaker-independent speech databases. To facilitate sharing between models, we used deleted interpolation to combine robust models with detailed ones. By combining poorly trained (context-dependent, generalized context, function-word-dependent, speaker-dependent) models with well-trained (context-independent, speaker-independent, uniform) models, we improved trainability through sharing.

To improve specificity, we used multiple codebooks of various LPC-derived features, and integrated external knowledge sources into the system.

We also improved the phone set to include multiple representations of some phones. We introduced the use of function-word-dependent phone modeling and generalized triphone modeling. Finally, we experimented with speaker adaptation algorithms based on speaker cluster selection and interpolated re-estimation.

9.2. Contributions

The most significant contribution of this work is the demonstration of a highly accurate large-vocabulary speaker-independent continuous speech recognizer. We have shown that full utilization of plentiful training can compensate for the deficiencies of speaker-independent training. However, better results did not simply materialize from adding training data. Instead, the modeling techniques had to be improved to take advantage of the ample training data. Our improvements were focused on four major problems with large-vocabulary speaker-independent continuous speech recognition:

- ***Finding a good model of speech***—We need a model of speech that can account for a large number of parameters, and that can be efficiently trained with considerable training material.
    - We adopted the currently popular hidden Markov modeling as the basis of SPHINX.
    - We described evaluation, recognition, and training algorithms for HMMs.
    - We described and experimented with various implementational issues in a baseline system:
        - signal processing,
        - vector quantization,
        - choice of model and dictionary,
        - smoothing of the output pdf's,
        - logarithmic representation of probabilities.
- ***Adding human knowledge***—Current systems, particularly statistical learning ones, have very little human knowledge. We investigated various ways of incorporating human knowledge in SPHINX:
    - Fixed-width parameters:
        - use of a bilinear transform for mel-scale warping;
        - use of differential coefficients to capture dynamic

information;

- use of power and differential power;
- investigation of various ways to integrate these fixed-width parameters. We found that multiple codebooks led to substantial improvements in recognition accuracy.

- Variable-width parameters:
  - addition of duration was very helpful for simple HMMs when no grammar was used;
  - addition of a network with phonetic probabilities produced using complex acoustic-phonetic features was *not* helpful, because the acoustic-phonetic network was not sufficiently accurate, and the use of segment-level integration was *ad-hoc*.
- Lexical-phonetic knowledge and tuning:
  - simple phonological rules;
  - implicit probabilistic insertion/deletion modeling;
  - modifications to the set of phones, and finding a good HMM topology for each phone;
  - these modifications improved SPHINX's recognition accuracy considerably.

- ***Finding a good unit of speech***—In order to model continuous speech, we need a speech unit that is well-understood, trainable (either frequent or sharable), and takes into account contextual dependencies.
  - We examined some of the currently used units:
    - word models are untrainable for large vocabularies;
    - phone models are trainable, but ignore contextual effects;
    - multi-phone units and transition models result in too many parameters that do not facilitate sharing;
    - word-dependent phone modeling is better than word modeling, but still uses too many models, and is not easily extendible to new or large tasks;
    - context-dependent phone modeling works quite well; however, it uses too many models and does not account for contexts that are similar.

- We introduced two new units:
  - function-word-dependent phone modeling—use of word-dependent phone modeling for function words only;
  - generalized triphone modeling—automatically learn similar contexts using an information theoretic clustering algorithm. These new context-sensitive phones are then used for training.
- Function-word-dependent phone modeling improved performance significantly by adding a small number of new models.
- Generalized triphone modeling improved recognition substantially, and is better than context-dependent phone modeling.
- Combining function-word-dependent phone models and generalized triphone models led to additional improvements.
- We used *deleted interpolation* to automatically estimate the interpolation weights of different estimates, based on how well each estimate predicts new data.

- ***Speaker learning and adaptation***—Although very good results were obtained for speaker-independent recognition, better results should be obtainable with some knowledge about a speaker. We investigated two speaker adaptation techniques:
  - Speaker cluster selection—
    - divide training speakers into speaker clusters;
    - generate codebook and HMMs for each cluster;
    - before recognizing a new speaker, first identify which cluster is the most appropriate;
    - use cluster-specific parameters (either directly or through probabilistic mapping) for recognition.
  - Interpolated re-estimation—Use deleted interpolation to combine various estimates into speaker adapted parameters:
    - speaker-independent parameters;
    - speaker-dependent parameters;
    - speaker-dependent parameters, with tied phone states;

  - co-occurrence smoothed speaker-dependent parameters;
  - similar-speaker parameters, from the most similar training speakers.
- Interpolated re-estimation led to modest improvements, while speaker cluster selection resulted in no improvement.

9.3. Future Work

Hidden Markov modeling has several very important properties that contributed to the success of SPHINX. The ability of HMMs to automatically optimize parameters from data is extremely powerful, the HMM integrated search that considers all of the knowledge sources at every step is very effective, and the absorption of faulty structural assumptions is most forgiving. By turning an *unknown structure problem* into an *unknown parameter problem*, and by automatically optimizing these parameters, HMM and maximum likelihood estimation are one of the most powerful learning paradigms available today. However, hidden Markov models have a number of serious problems, some of which are due to our implementational choices and others to the nature of HMMs. These problems are very challenging, and are areas of continuing research.

The first problem is our use of vector quantization. Our choice to use discrete HMMs with VQ was dictated by expedience. Discrete HMMs enabled us to perform many more experiments than continuous HMMs would have permitted. Moreover, continuous densities are not as well understood, and there are still many disagreements and problems regarding their use. However, we realize that ultimately vector quantization is undesirable, and one area of future research is to directly model the speech parameters with continuous density HMMs.

Another problem with our implementation is the use of Viterbi search. While Viterbi is easy to implement, and led to very good results for moderately large tasks, it finds the optimal state sequence rather than the optimal word sequence. In the future, we would like to explore tree-based search decoding procedures that try to find the optimal word sequence.

The above problems are related to our implementational choices, and can be solved within the framework of hidden Markov models and maximum likelihood estimation (MLE). However, there are also fundamental problems with HMM and MLE. First, the Markov assumption and output independence assumption are inaccurate. Since HMMs are generative models, it would be revealing to consider the speech synthesis process that HMMs assume. A discrete HMM synthesizer for a phoneme consists of states, transitions, and a number of centisecond speech templates, corresponding to the VQ codewords. At every centisecond, two many-faced coins are flipped. The first coin determines the next state, and the second coin selects a centisecond template according to the output pdf. This template is synthesized for one centisecond, and the process continues without any memory of the past. This is a very inaccurate model of the speech production process. However, adding memory to HMMs is very costly, because the number of parameters increases exponentially for n^{th}-order HMMs. There are simple ways of compensating for this problem, such as using differential coefficients to give each frame more scope, or considering the correlation between adjacent frames [Brown 87]. However, they cannot completely solve the problem because a great amount of memory of the past is needed for some speech phenomena. Moreover, we certainly do not produce speech in fixed-width frames—sequence-modeling is more perceptually plausible. But to model sequences usually requires segmentation, which cannot be done reliably, and hidden Markov modeling of variable-width events (such as phonemes) always led to inferior performance [Bahl 78b, Nag 86]. A different approach to sequence modeling is the *stochastic segment model* [Roucos 87, Roucos 88]. While this is a promising approach, they have yet to demonstrate consistently superior performance over HMMs.

Another problem is the use of maximum likelihood estimation. The object of speech recognition is to *discriminate* among the words in the lexicon. MLE does not make any effort to that end, yet it has worked surprisingly well for HMMs. Even more perplexing is the fact that attempts to discriminate among words did not lead to superior results for larger vocabularies. For example, Cole *et al.* [Cole 83] used a hand-optimized decision tree to discriminate letters of the alphabet. This technique worked well for that task, but did not work well for large-vocabulary word recognition. Lee [Lee 86] used subword-level clustering to isolate discriminating components of the lexicon. Again, it worked well for the task, but not for large vocabulary. Brown [Brown 87] introduced maximum mutual information estimation (MMIE) for HMMs, which produced very good results for E-Set recognition; however, MMIE has not yet outperformed MLE for large vocabularies. Thus, we believe that for large vocabularies, discrimination is more difficult. Ideally, a recognizer should only consider a small set of *cohorts* [Shipman 82] for each possible word. The question is, then, how to discriminate between confusable words, and ignore the "noise"?

One other problem concerns using subword models. We have shown that using more models lead to better results, at the cost of potential estimation problems. We have used interpolation to allow sharing of training data among models. But it would be even better to have sufficient training data for the most detailed models. The way to achieve this is to record more training data. We are currently planning to enlarge our database of speech, from which we hope to build a very-large-vocabulary recognizer. Also, recent studies [Bahl 88b, LeeCH 88b] reported high accuracies with *acoustic* subword models. These acoustic subword models are defined using measures of acoustic similarity, which may or may not bear resemblance to phonetically meaningful units. The choice of acoustic subword models is another fascinating problem.

Finally, our results with speaker adaptation have not reached our original expectations. We attribute this to the robustness and the high accuracy of speaker-independent SPHINX, as well as the lack of training and adaptation data for our techniques. In the future, we would like to use more training material and test our hypothesis, as well as examine other more rapid adaptation procedures.

Therefore, while our results are very encouraging, we must not be complacent: there are many difficult challenges that we have not begun to address. These problems will probably require decades of research before any solutions can be found. We look forward to continuing research in these areas.

9.4. Final Remarks

Speaker-independent speech recognition has lost some popularity recently due to the disappointing results of some systems. Many researchers have become more skeptical about speaker-independent recognition, and even more skeptical when the task is large-vocabulary continuous speech.

Although we have always been convinced of the utility and necessity of speaker-independent large-vocabulary continuous speech recognition, we were not overly optimistic about the prospects of a highly accurate recognizer that many veterans of speech recognition have shunned. It was originally hoped that an acceptable accuracy would be reached in this study, and that some lessons could be learned in the process.

Indeed, we have learned many lessons. The most important and pleasantly surprising lesson is that *large-vocabulary speaker-independent continuous speech recognition is feasible.* This is made possible by another lesson, namely, *with a powerful learning paradigm, the performance of a system can always be improved with more training data, subject to our ability to make the models more sophisticated.* We have also learned that *improvements can come from not only better modeling techniques, but also speech knowledge and careful analysis of the task and system errors.*

This study introduced a number of new concepts, and the principles behind each concept. These improvements have reduced the error rate of our baseline system by 85%, resulting in the most accurate large-vocabulary speaker-independent continuous speech recognizer today. We have not, by any means, solved the speech recognition problem, but it is our hope that our research has provided some insight about statistical learning, dispelled some cynicism about speaker-independent recognition, and will help to point the way for future research.

Appendix I
Evaluating Speech Recognizers

I.1. Perplexity

The performance of a speech recognizer is a function of several variables:

- The quality of acoustic modeling.
- The quality of language modeling.
- The constraint imposed by the grammar (if any).
- The inherent confusability of the vocabulary.

Throughout this thesis, we have been concerned with improved acoustic modeling. However, in order to compare different systems or different language models, the other three factors must be taken into account. *Perplexity* is a measure of the constraint imposed by the grammar, or the level of uncertainty given the grammar.

Before we define *perplexity*, let's first consider how a grammar reduces uncertainty during recognition. Without a grammar, the entire vocabulary must be considered at every decision point. With a grammar, it is possible to eliminate many candidates from consideration, or to assign higher probabilities to some candidates than others. This constraint at a decision point (j) can be measured by *entropy* (H), or the number of bits necessary to specify the next word using an optimal encoding scheme:

$$H(W \mid j) = -\sum_{w=1}^{V} P(w \mid j) \cdot \log_2 [P(w \mid j)] \qquad (1)$$

The perplexity at the decision point j is defined to be:

$$Q(w \mid j) = 2^{H(w \mid j)} \qquad (2)$$

If we have a finite state grammar with many states, or decision points, then entropy is computed as:

$$H(L) = \sum_{j} \pi(j) \, H(W \mid j) \qquad (3)$$

where $\pi(j)$ is the steady-state probability of being in state j. The per-word perplexity of this language is:

$$Q(L) = 2^{H(L)} \tag{4}$$

The above method for computing entropy and perplexity are useful when the *true language model* is a finite state grammar with or without probabilities. But in some cases (such as IBM's natural language task), the true language model is very different from the trigram model used, and its perplexity cannot be measured. In other words, the perplexity measured from the language model does not reflect the uncertainty encountered during recognition. When this is the case, *test-set perplexity* [Jelinek 85, Kimball 86] should be used. Test-set perplexity is simply the geometric mean of probabilities at each decision point for the test set sentences, or:

$$\frac{1}{n}\log P(w_1,w_2, \cdots ,w_n) \tag{5}$$

where $P(w_1,w_2, \cdots ,w_n)$ is the probability of generating a string of n words. These n words are the concatenation of many sentences, with an end-of-sentence counting as one word. In the case of a bigram grammar, for one sentence:

$$P(w_1,w_2, \cdots ,w_m) \approx \\ P(w_1 \mid Sent\text{–}begin) \cdot P(w_2 \mid w_1) \cdot \cdots \cdot P(Sent\text{–}end \mid w_m) \tag{6}$$

For tasks whose true language models are known, it can be shown that as the number of test sentences approaches infinity, test-set perplexity (Equation 5) is equivalent to perplexity (Equation 4). However, for tasks whose true language models are not known, test-set perplexity will be higher than perplexity measured from the language model because it is inaccurate. In our word-pair and bigram grammars, the test-set perplexity should be the same as the perplexity measured on the finite state grammars, because the true language models are known. The word-pair grammar has a perplexity of about 60 (which makes sense because there are 57,878 word pairs and 997 words, and 60 is about $\frac{57878}{997}$), and the bigram has a perplexity of about 20.

I.2. Computing Error Rate

For isolated-word recognition, computing the error rate is straightforward because the only possible type of error is substitution (an incorrect word was substituted for the correct word). However, in continuous speech, there are three types of errors: substitution, deletion (a correct word was omitted in the recognized sentence), and insertion (an extra word was added in the recognized sentence). Clearly, substitutions and deletions are errors. But it is not clear whether insertions should be counted as errors. On the one hand, a word not expected was inserted. This is clearly undesirable, and should be penalized. On the other hand, it is not really fair to consider (*recognize* → *wreck a nice*) as three errors. Earlier work tended not to count insertions as errors, while more recent ones counted them as errors. In order to enable comparison against all systems, and because of the doubtful nature of insertion errors, we report both results : *percent correct*, which doesn't consider insertions as errors, and *word accuracy*, which does.

To determine the recognition accuracy, we first align the recognized word string against the correct word string, and then compute the number of words *Correct, Substitutions, Deletions, Insertions*. This alignment can be obtained using a dynamic programming algorithm.[18] Finally, *Percent Correct* and *Word Accuracy* are computed by:

$$Percent\ Correct = 100 \cdot \frac{Correct}{Correct\ Sent\ Length} \tag{7}$$

$$Error\ Rate = 100 \cdot \frac{Subs + Dels + Ins}{Correct\ Sent\ Length} \tag{8}$$

$$Word\ Accuracy = 1 - Error\ Rate$$

Since $Correct\ Sent\ Length = Correct + Subs + Dels$,

$$Word\ Accuracy = 100 \cdot \frac{Correct - Ins}{Correct\ Sent\ Length} \tag{9}$$

Therefore, *Percent Correct* and *Word Accuracy* differ by the number of insertions.

Confusions between homonyms are considered correct recognitions when no language model is used, because homonyms have the identical dictionary entries, and are indistinguishable. When a grammar is used, homonym confusions are counted as substitution errors.

[18]The actual dynamic programming program was provided by the National Bureau of Standards.

Appendix II
The Resource Management Task

II.1. The Vocabulary and the SPHINX Pronunciation Dictionary

A	AX
A42128	EY F AO R T UW W AH N T UW EY TD
AAW	EY EY D AH B AX Y UW
ABERDEEN	AE B ER D IY N
ABOARD	AX B AO R DD
ABOVE	AX B AH V
ADD	AE DD
ADDED	AE DX IX DD
ADDING	AE DX IX NG
AFFECT	AX F EH K TD
AFTER	AE F T ER
AGAIN	AX G EH N
AJAX	EY JH AE K S
AJAX'S	EY JH AE K S IX Z
ALASKA	AX L AE S K AX
ALERT	AX L ER TD
ALERTS	AX L ER TS
ALEXANDRIA	AE L IX G Z AE N D R IY AX
ALL	AA L
AN	AX N
ANCHORAGE	AE NG K R IH JH
AND	EH N DD
ANY	EH N IY
ANYBODY	EH N IY B AH DX IY
APALACHICOLA	AE P AX L AE CH IX K OW L AX
APALACHICOLA'S	AE P AX L AE CH IX K OW L AX Z
APRIL	EY P R L
ARABIAN	AX R EY B IY IX N
ARCTIC	AA R KD T IX KD
ARE	AA R
AREA	EH R IY AX
AREAS	EH R IY AX Z
AREN'T	AA R N TD
ARKANSAS	AA R K AX N S AO
ARKANSAS'S	AA R K AX N S AO Z
AROUND	AX R AW N DD
ARRIVAL	AX R AY V L
ARRIVE	AX R AY V
ARRIVED	AX R AY V DD
ARRIVING	AX R AY V IX NG

ARROW	AE R OW
AS	EH Z
ASTORIA	AE S D AO R IY AX
ASUW	EY EH S Y UW D AH B AX Y UW
ASW	EY EH S D AH B AX Y UW
AT	EH TD
ATLANTIC	AE TD L AE N IX KD
AUGUST	AA G AX S TD
AVAILABLE	AX V EY L AX B L
AVERAGE	AE V AX R IX JH
BAD	B AE DD
BADGER	B AE JH ER
BADGER'S	B AE JH ER Z
BAINBRIDGE	B EY N B R IH JH
BAINBRIDGE'S	B EY N B R IH JH IX Z
BANGKOK	B AE NG K AA KD
BARGE	B AA R JH
BASS	B AE S
BAY	B EY
BE	B IY
BEAM	B IY M
BEAMS	B IY M Z
BEEN	B IH N
BEFORE	B IX F AO R
BELOW	B IX L OW
BERING	B EH R IX NG
BETTER	B EH DX ER
BETWEEN	B IH T W IY N
BIDDLE	B IH DX L
BIDDLE'S	B IH DX L Z
BISMARK	B IH Z M AA R KD
BOMBAY	B AA M B EY
BOTH	B OW TH
BOX	B AA K S
BRIGHT	B R AY TD
BRITISH	B R IH DX IX SH
BROOKE	B R UH KD
BROOKE'S	B R UH K S
BRUNSWICK	B R AH N Z W IH KD
BRUNSWICK'S	B R AH N Z W IH K S
BUD-TEST	B AH T EH S TD
BUMP	B AH M PD
BY	B AY
C-CODE	S IY K OW DD
C-CODES	S IY K OW D Z
C-RATING	S IY R EY DX IX NG
C-RATINGS	S IY R EY DX IX NG Z
C1	S IY W AH N

C2	S IY T UW
C3	S IY TH R IY
C4	S IY F AO R
C5	S IY F AY V
CALIFORNIA	K AE L AX F AO R N Y AX
CAMDEN	K AE M D IX N
CAMDEN'S	K AE M D IX N Z
CAMPBELL	K AE M B L
CAMPBELL'S	K AE M B L Z
CAN	K IX N
CANADA	K AE N AX DX AX
CAPABILITIES	K EY P AX B IH L AX DX IY Z
CAPABILITY	K EY P AX B IH L AX DX IY
CAPABLE	K EY P AX B L
CAPACITIES	K AX P AE S IX DX IY Z
CAPACITY	K AX P AE S IX DX IY
CARRIER	K EH R IY ER
CARRIER'S	K EH R IY ER Z
CARRIERS	K EH R IY ER Z
CARRIERS'S	K EH R IY ER Z
CARRY	K AE R IY
CASREP	K AE Z R EH PD
CASREPED	K AE Z R EH P DD
CASREPS	K AE Z R EH P S
CASUALTY	K AE SH L DX IY
CAT-2	K AE T UW
CAT-3	K AE TH R IY
CAT-4	K AE TD F AO R
CATEGORIES	K AE DX AX G AO R IY Z
CATEGORY	K AE DX AX G AO R IY
CENTER	S EH N ER
CENTERED	S EH N ER DD
CENTERING	S EH N ER IX NG
CEP	S IY IY P IY
CHANGE	CH EY N JH
CHANGED	CH EY N JH DD
CHANGING	CH EY N JH IX NG
CHANNEL	CH AE N L
CHART	CH AA R TD
CHARTS	CH AA R TS
CHATTAHOOCHEE	CH AE DX AX HH UW CH IY
CHATTAHOOCHEE'S	CH AE DX AX HH UW CH IY Z
CHESHIRE	CH EH SH ER
CHINA	CH AY N AX
CHOP	CH AA PD
CHOPPED	CH AA P TD
CHOPPING	CH AA P IX NG
CITRUS	S IH T R AX S

CITRUS'S	S IH T R AX S IX S
CITY	S IH DX IY
CLEAR	K L IH R
CLEARED	K L IH R DD
CLEARING	K L IH R IX NG
CLEVELAND	K L IY V L AX N DD
CLEVELAND'S	K L IY V L AX N D Z
CLOSE	K L OW S
CLOSER	K L OW S ER
CLOSEST	K L OW S IX S TD
CODAG	K OW D AE G
CODE	K OW DD
CODES	K OW D Z
COLOR	K AH L ER
COMPARED	K AX M P EH R DD
CONFIDENCE	K AA N F IX DX EH N S
CONFIDENCE'S	K AA N F IX DX AX N S IX Z
CONIFER	K AA N AX F ER
CONIFER'S	K AA N AX F ER Z
CONQUEST	K AA N K W EH S TD
CONQUEST'S	K AA N K W EH S TS
CONSTANT	K AA N S T AX N TD
CONSTANT'S	K AA N S T AX N TS
CONSTELLATION	K AA N S T AX L EY SH AX N
CONSTELLATION'S	K AA N S T AX L EY SH AX N Z
CONVENTIONAL	K AX N V EH N SH AX N L
COOK	K UH KD
COPELAND	K OW P L AX N DD
COPELAND'S	K OW P L AX N D Z
CORAL	K AO R L
COULD	K UH DD
COULDN'T	K UH D IX N TD
COUNT	K AW N TD
COUNTED	K AW N IX DD
COUNTING	K AW N IX NG
CROVL	K R OW V L
CROVLS	K R OW V L Z
CRUISER	K R UW Z ER
CRUISER'S	K R UW Z ER Z
CRUISERS	K R UW Z ER Z
CRUISERS'S	K R UW Z ER Z
CRUISING	K R UW Z IX NG
CURRENT	K ER AX N TD
CURRENTLY	K ER AX N TD L IY
DALE	D EY L
DALE'S	D EY L Z
DARWIN	D AA R W IX N
DATA	D EY DX AH

DATE	D EY TD
DATED	D EY DX IX DD
DATES	D EY TS
DAVIDSON	D EY V IX DD S AX N
DAVIDSON'S	D EY V IX DD S AX N Z
DAY	D EY
DAYS	D EY Z
DDD992	D IY D IY D IY N AY N N AY N T UW
DECEMBER	D IX S EH M B ER
DECREASE	D IY K R IY S
DECREASED	D IY K R IY S TD
DECREASING	D IY K R IY S IX NG
DEFAULT	D IX F AO L TD
DEFAULTS	D IX F AO L TS
DEFINE	D IX F AY N
DEFINED	D IX F AY N DD
DEFINING	D IX F AY N IX NG
DEFINITION	D EH F IX N IH SH IX N
DEFINITIONS	D EH F IX N IH SH IX N Z
DEGRADATION	D EH G R AX D EY SH IX N
DEGRADATIONS	D EH G R AX D EY SH IX N Z
DEGRADE	D IY G R EY DD
DEGRADED	D IY G R EY DX IX DD
DEGRADING	D IY G R EY DX IX NG
DEGREES	D IX G R IY Z
DELETE	D IX L IY TD
DELETED	D IX L IY DX IX DD
DELETING	D IX L IY DX IX NG
DENVER	D EH N V ER
DENVER'S	D EH N V ER Z
DEPLOYED	D IX P L OY DD
DEPLOYMENT	D IX P L OY M AX N TD
DEPLOYMENTS	D IX P L OY M AX N TS
DEPTH	D EH P TH
DEPTHS	D EH P TH S
DESTINATION	D EH S T AX N EY SH IX N
DESTINATIONS	D EH S T AX N EY SH IX N Z
DID	D IH DD
DIDN'T	D IH DD EN TD
DIEGO-GARCIA	D IY EY G OW G AA R S IY AX
DIESEL	D IY Z L
DIM	D IH M
DISPLACEMENT	D IH S B L EY S M AX N TD
DISPLACEMENTS	D IH S B L EY S M AX N TS
DISPLAY	D IH S B L EY
DISPLAYED	D IH S B L EY DD
DISPLAYING	D IH S B L EY IX NG
DISTANCE	D IH S T IX N S

DIXON	D IH K S IX N
DIXON'S	D IH K S IX N Z
DMDS	D IY EH M D IY EH S
DO	D UW
DOES	D AH Z
DOESN'T	D AH Z AX N TD
DON'T	D OW N TD
DOWNES	D AW N Z
DOWNES'S	D AW N Z IX Z
DOWNGRADE	D AW N G R EY DD
DOWNGRADED	D AW N G R EY DX IX DD
DRAFT	D R AE F TD
DRAFTS	D R AE F TS
DRAW	D R AO
DUBUQUE	D AX B Y UW KD
DUBUQUE'S	D AX B Y UW K S
DUE	D UW
DURING	D ER IX NG
EACH	IY CH
EARLIER	ER L IY ER
EARLIEST	ER L IY IX S TD
EARLY	ER L IY
EAST	IY S TD
EASTERN	IY S T ER N
EASTPAC	IY S TD P AE K
EASTPAC'S	IY S TD P AE K S
ECG041	IY S IY JH IY Z IY R OW F AO R W AH N
ECHO	EH K OW
ECONOMIC	EH K IX N AA M IX KD
EDIT	EH DX IH TD
EDITED	EH DX IH DX IX DD
EDITING	EH DX IH DX IX NG
EIGHT	EY TD
EIGHTEEN	EY T IY N
EIGHTEENTH	EY T IY N TH
EIGHTH	EY TH
EIGHTY	EY D IY
EISENHOWER	AY Z IX N HH AW ER
EISENHOWER'S	AY Z IX N HH AW ER Z
ELEVEN	AX L EH V IH N
ELEVENTH	AX L EH V IH N TH
EMPLOYED	EH M P L OY DD
END	EH N DD
ENDING	EH N D IX NG
ENGLAND	IH NG G L AX N DD
ENGLAND'S	IH NG G L AX N D Z
ENGLISH	IH NG G L IH SH

ENOUGH	IH N AH F
ENROUTE	EH N R UW TD
ENTERPRISE	EH N ER P R AY Z
ENTERPRISE'S	EH N ER P R AY Z IX Z
EQUIPMENT	IX K W IH PD M AX N TD
EQUIPPED	IX K W IH P TD
ESTEEM	EH S D IY M
ESTEEM'S	EH S D IY M Z
ESTIMATED	EH S T AX M EY DX IX DD
ETA	IY T IY EY
ETR	IY T IY AA R
EVER	EH V ER
EVERETT	EH V AX R IH TD
EXPECTED	EH K S B EH K T IX DD
FANNING	F AE N IX NG
FANNING'S	F AE N IX NG Z
FAR	F AA R
FARTHER	F AA R DH ER
FARTHEST	F AA R DH AX S TD
FAST	F AE S TD
FASTER	F AE S T ER
FASTEST	F AE S T IX S TD
FEBRUARY	F EH B Y UW EH R IY
FEET	F IY TD
FFF088	EH F EH F EH F Z IY R OW EY TD EY TD
FIFTEEN	F IH F T IY N
FIFTEENTH	F IH F T IY N TH
FIFTH	F IH F TH
FIFTY	F IH F T IY
FIGURE	F IH G ER
FIGURES	F IH G ER Z
FIJI	F IY JH IY
FIND	F AY N DD
FIREBUSH	F AY R B UH SH
FIREBUSH'S	F AY R B UH SH IX Z
FIRST	F ER S TD
FIVE	F AY V
FIXED	F IH K S TD
FLASHER	F L AE SH ER
FLASHER'S	F L AE SH ER Z
FLEET	F L IY TD
FLEETS	F L IY TS
FLINT	F L IH N TD
FLINT'S	F L IH N TS
FOOTER	F UH DX ER
FOR	F ER
FORMOSA	F ER M OW S AX
FORTY	F AO R DX IY

FOUR	F AO R
FOURTEEN	F AO R T IY N
FOURTEENTH	F AO R T IY N TH
FOURTH	F AO R TH
FOX	F AA K S
FOX'S	F AA K S IX Z
FREDERICK	F R EH D R IH KD
FREDERICK'S	F R EH D R IH K S
FRENCH-POLYNESIA	F R EH N CH P AA L AX N IY SH AX
FRESNO	F R EH Z N OW
FRESNO'S	F R EH Z N OW Z
FRIDAY	F R AY DX EY
FRIDAY'S	F R AY DX EY Z
FRIGATE	F R IH G IH TD
FRIGATE'S	F R IH G IH TS
FRIGATES	F R IH G IH TS
FRIGATES'S	F R IH G IH TS
FROM	F R AX M
FUEL	F Y UW L
FULL	F UH L
GALLONS	G AE L AX N Z
GALVESTON	G AE L V AX S T AX N
GAS	G AE S
GET	G EH TD
GIVE	G IH V
GLACIER	G L EY SH ER
GLACIER'S	G L EY SH ER Z
GNOMONIC	N OW M AA N IH KD
GO	G OW
GOING	G OW IX NG
GONE	G AA N
GREAT	G R EY TD
GREAT-CIRCLE	G R EY TD S ER K L
GREATER	G R EY DX ER
GREATEST	G R EY DX IX S TD
GREEN	G R IY N
GRID	G R IH DD
GRIDLEY	G R IH DD L IY
GRIDLEY'S	G R IH DD L IY Z
GRILL	G R IH L
GROSS	G R OW S
GROUP	G R UW PD
GROUPS	G R UW P S
GUARDFISH	G AA R DD F IH SH
GUARDFISH'S	G AA R DD F IH SH IX Z
GUITARRO	G IH T AA R OW
GUITARRO'S	G IH T AA R OW Z
GULF	G AA L F

HAD	HH AE DD
HALF	HH AE F
HARBOR	HH AA R B ER
HARPOON	HH AA R P UW N
HAS	HH AE Z
HASN'T	HH AE Z IX N TD
HAVE	HH AE V
HAVEN'T	HH AE V AX N TD
HAWKBILL	HH AA KD B IH L
HAWKBILL'S	HH AA KD B IH L Z
HE	HH IY
HE'S	HH IY Z
HECTOR	HH EH KD T ER
HECTOR'S	HH EH KD T ER Z
HEPBURN	HH EH PD B ER N
HEPBURN'S	HH EH PD B ER N Z
HER	HH ER
HERS	HH ER Z
HFDF	EY CH EH F D IY EH F
HIGH	HH AY
HIGHER	HH AY ER
HIGHEST	HH AY IX S TD
HIM	HH IH M
HIS	HH IH Z
HOME	HH OW M
HOMER	HH OW M ER
HONG-KONG	HH AA NG K AA NG
HONOLULU	HH AA N AH L UW L UW
HOOKED	HH UH K TD
HORNE	HH AO R N
HORNE'S	HH AO R N Z
HOUR	AW ER
HOURS	AW ER Z
HOW	HH AW
HUNDRED	HH AH N D R AX DD
ICE-NINE	AY S N AY N
ID	AY D IY
IDENTIFICATION	AY D EH N IH F AX K EY SH IX N
IDENTIFICATIONS	AY D EH N IH F AX K EY SH IX N Z
IF	IH F
IN	IX N
INCLUDE	IH N K L UW DD
INCLUDED	IH N K L UW DX IX DD
INCLUDING	IH N K L UW DX IX NG
INCREASE	IH N K R IY S
INCREASED	IH N K R IY S TD
INCREASING	IH N K R IY S IX NG
INDEPENDENCE	IH N D AX P EH N D IX N S

INDEPENDENCE'S	IH N D AX P EH N D IX N S IX Z
INDIAN	IH N D IY IX N
INDONESIA	IH N D AX N IY SH AX
INFORMATION	IH N F ER M EY SH IX N
INSTALLED	IH N S D AA L DD
INSTEAD	IH N S D EH DD
INSUFFICIENT	IH N S AX F IH SH AX N TD
INVOLVED	IH N V AA L V DD
INVOLVING	IH N V AA L V IX NG
IRONWOOD	AY R N W UH DD
IRONWOOD'S	AY R N W UH D Z
IS	IH Z
ISLANDS	AY L AX N D Z
ISN'T	IH Z EN TD
IT	IH TD
IT'S	IH TS
ITS	IH TS
JANUARY	JH AE N Y UW EH R IY
JAPAN	JH AX P AE N
JARRETT	JH EH R IX TD
JARRETT'S	JH EH R IX TS
JARVIS	JH AA R V IX S
JARVIS'S	JH AA R V IX S IX Z
JASON	JH EY S IX N
JASON'S	JH EY S IX N Z
JULY	JH AX L AY
JUNE	JH UW N
JUPITER	JH UW P AX DX ER
JUPITER'S	JH UW P AX DX ER Z
KENNEDY	K EH N AX DX IY
KENNEDY'S	K EH N AX DX IY Z
KILOMETER	K AX L AA M AX DX ER
KILOMETERS	K AX L AA M AX DX ER Z
KIRK	K ER KD
KIRK'S	K ER K S
KISKA	K IH S K AX
KISKA'S	K IH S K AX Z
KNOT	N AA TD
KNOTS	N AA TS
KODIAK	K OW DX IY AE KD
KOREA	K ER IY AX
KOREAN	K ER IY AX N
LAMPS	L AE M P S
LANTFLT	L AE N TD F L IY TD
LARGE	L AA R JH
LARGER	L AA R JH ER
LARGEST	L AA R JH IX S TD
LAST	L AE S TD

LAT	L AE TD
LAT-LON	L AE TD L AO N
LAT-LONS	L AE TD L AO N Z
LATER	L EY DX ER
LATEST	L EY DX IX S TD
LATITUDE	L AE DX IH T UW DD
LATITUDES	L AE DX IH T UW D Z
LATS	L AE TS
LEAST	L IY S TD
LEFT	L EH F TD
LENGTH	L EH NG TH
LENGTHS	L EH NG TH S
LENINGRAD	L EH N IH N G R AE DD
LESS	L EH S
LETTER	L EH DX ER
LETTERS	L EH DX ER Z
LEVEL	L EH V L
LEVELS	L EH V L Z
LINK-11	L IH NG K AX L EH V AX N
LIST	L IH S TD
LOCATED	L OW K EY DX IX DD
LOCATION	L OW K EY SH IX N
LOCATIONS	L OW K EY SH IX N Z
LOCKWOOD	L AA K W UH DD
LOCKWOOD'S	L AA K W UH D Z
LON	L AA N
LONG	L AA NG
LONGER	L AA NG G ER
LONGEST	L AA NG G AX S TD
LONGITUDE	L AA N JH IX T UW DD
LONGITUDES	L AA N JH IX T UW D Z
LONS	L AA N Z
LOW	L OW
LOWER	L OW ER
LOWEST	L OW IX S TD
M-CODE	EH M K OW DD
M-CODES	EH M K OW D Z
M-RATING	EH M R EY DX IX NG
M-RATINGS	EH M R EY DX IX NG Z
M1	EH M W AH N
M2	EH M T UW
M3	EH M TH R IY
M4	EH M F AO R
M5	EH M F AY V
MADAGASCAR	M AE DX AX G AE S K ER
MADE	M EY DD
MAKE	M EY KD
MAKING	M EY K IX NG

MANCHESTER	M AE N CH EH S T ER
MANHATTAN	M AE N HH AE TD EN
MANHATTAN'S	M AE N HH AE TD EN Z
MANILA	M AX N IH L AX
MANY	M EH N IY
MARCH	M AA R CH
MARS	M AA R S
MARS'S	M AA R Z IX Z
MAX	M AE K S
MAXIMUM	M AE K S IX M AX M
MAY	M EY
MCCLUSKY	M AX K L AH S K IY
MCCLUSKY'S	M AX K L AH S K IY Z
ME	M IY
MEASURE	M EH SH ER
MERCATOR	M ER K EY DX ER
MERCURY	M ER K Y ER IY
MERCURY'S	M ER K Y ER IY Z
METEOR	M IY DX IY ER
METEOR'S	M IY DX IY ER Z
METERS	M IY DX ER Z
METRIC	M EH T R IX KD
MEXICO	M EH K S IX K OW
MIAMI	M AY AE M IY
MIDGETT	M IH JH IX TD
MIDGETT'S	M IH JH IX TS
MIDPAC	M IH DD P AE KD
MIDPAC'S	M IH DD P AE K S
MIDWAY	M IH DD W EY
MIDWAY'S	M IH DD W EY Z
MILE	M AY L
MILES	M AY L Z
MIND	M AY N DD
MINUTE	M IH N AX TD
MINUTES	M IH N AX TS
MISHAWAKA	M IH SH AX W AA K AH
MISHAWAKA'S	M IH SH AX W AA K AH Z
MISSION	M IH SH AX N
MISSIONS	M IH SH AX N Z
MISSISSIPPI	M IH S IX S IH P IY
MISSISSIPPI'S	M IH S IX S IH P IY Z
MIW	EH M AY D AH B AX Y UW
MOB	EH M OW B IY
MONDAY	M AH N D EY
MONDAY'S	M AH N D EY Z
MONGOLIA	M AA N G OW L IY AX
MONTH	M AH N TH
MONTH'S	M AH N TH S

MONTHS	M AH N TH S
MONTICELLO	M AA N IX S EH L OW
MONTICELLO'S	M AA N IX S EH L OW Z
MORE	M AO R
MOST	M OW S TD
MOZAMBIQUE	M OW Z AE M B IY KD
MUCH	M AH CH
N92762	EH N N AY N T UW S EH V AX N S IH K S T UW
NAME	N EY M
NAMES	N EY M Z
NAPLES	N EY P L Z
NASHUA	N AE SH UW AX
NASHUA'S	N AE SH UW AX Z
NEAR	N IY R
NEARER	N IY R ER
NEAREST	N IY R AX S TD
NEVER	N EH V ER
NEW	N UW
NEW-CALEDONIA	N UW K AE L AX D OW N Y AX
NEW-YORK	N UW Y AO R KD
NEW-ZEALAND	N UW Z IY L AX N DD
NEWCASTLE	N UW K AE S L
NEWER	N UW ER
NEWEST	N UW AX S TD
NEXT	N EH K S TD
NINE	N AY N
NINETEEN	N AY N T IY N
NINETEENTH	N AY N T IY N TH
NINETY	N AY N DX IY
NINTH	N AY N TH
NO	N OW
NOME	N OW M
NORTH	N AO R TH
NORTHERN	N AO R DH ER N
NOT	N AA TD
NOVA	N OW V AX
NOVEMBER	N OW V EH M B ER
NOW	N AW
NTDS	EH N T IY D IY EH S
NUCLEAR	N UW K L IY ER
NUMBER	N AH M B ER
OAKLAND	OW K L AX N D
OCEAN	OW SH IX N
OCTOBER	AA KD T OW B ER
OF	AX V
OFF	AO F
OLD	AO L DD

OLDER	AO L D ER
OLDEST	AO L D AX S TD
OLYMPIA	OW L IH M P IY AX
ON	AA N
ONE	W AH N
ONLY	OW N L IY
OPEN	OW P AX N
OR	AO R
ORANGE	AO R AX N JH
OSGP	OW EH S JH IY P IY
OVERALL	OW V ER AA L
OVERLAY	OW V ER L EY
OVERLAYS	OW V ER L EY Z
PAC	P AE KD
PACFLT	P AE KD F L IY TD
PACIFIC	P AX S IH F IX KD
PANAMA	P AE N AX M AA
PARAMETER	P ER AE M AX DX ER
PARAMETERS	P ER AE M AX DX ER Z
PEARL-HARBOR	P ER L HH AA R B ER
PEORIA	P IY AO R IY AX
PEORIA'S	P IY AO R IY AX Z
PERCENT	P ER S EH N TD
PERSIAN	P ER SH EN
PERSONNEL	P ER S EN EH L
PHILIPPINE	F IH L AX P IY N
PHILIPPINES	F IH L AX P IY N Z
PIGEON	P IH JH IX N
PIGEON'S	P IH JH IX N Z
PLH003	P IY EH L EY CH Z IY R OW Z IY R OW TH R IY
PLUCK	P L AH KD
PLUCK'S	P L AH K S
PLUNGER	P L AH N JH ER
PLUNGER'S	P L AH N JH ER Z
POLLACK	P AA L AX KD
POLLACK'S	P AA L AX K S
PORT	P AO R TD
PORT-ELIZABETH	P AO R TD IY L IH Z AX B AX TH
PORT-VICTORIA	P AO R TD V IH KD T AO R IY AX
PORTS	P AO R TS
POSIT	P AA Z IX TD
POSITION	P AX Z IH SH AX N
POSITIONS	P AX Z IH SH AX N Z
POSITS	P AA Z IX TS
POUGHKEEPSIE	P AX K IH P S IY
POUGHKEEPSIE'S	P AX K IH P S IY Z
POWERED	P AW ER DD

PRAIRIE	P R EH R IY
PRAIRIE'S	P R EH R IY Z
PRESENT	P R EH Z AX N TD
PREVIOUS	P R IY V IY AX S
PROBLEM	P R AA B L AX M
PROBLEMS	P R AA B L AX M Z
PROJECTION	P R AX JH EH K SH AX N
PROPELLED	P R AX P EH L DD
PROPULSION	P R AX P AH L SH AX N
PUFFER	P AH F ER
PUFFER'S	P AH F ER Z
PUGET-1	P Y UW JH IH TD W AH N
QUARTER	K W AO R DX ER
QUARTERS	K W AO R DX ER Z
QUEENFISH	K W IY N F IH SH
QUEENFISH'S	K W IY N F IH SH IX Z
RADAR	R EY DX AA R
RAMSEY	R AE M Z IY
RAMSEY'S	R AE M Z IY Z
RANGER	R EY N JH ER
RANGER'S	R EY N JH ER Z
RATED	R EY DX IX DD
RATHBURNE	R AE TH B ER N
RATHBURNE'S	R AE TH B ER N Z
RATING	R EY DX IX NG
RATINGS	R EY DX IX NG Z
READINESS	R EH DX IY N EH S
REASON	R IY Z AX N
REASONER	R IY Z N ER
REASONER'S	R IY Z N ER Z
REASONS	R IY Z AX N Z
RECENT	R IY S EN TD
RECLAIMER	R IY K L EY M ER
RECLAIMER'S	R IY K L EY M ER Z
RED	R EH DD
REDEFINE	R IY D IX F AY N
REDEFINED	R IY D IX F AY N DD
REDEFINING	R IY D IX F AY N IX NG
REDISPLAY	R IY D IH S B L EY
REDO	R IY D UW
REDRAW	R IY D R AO
REEVES	R IY V Z
REEVES'S	R IY V Z IX Z
REMAINING	R IY M EY N IX NG
REMARK	R IY M AA R KD
REMARKS	R IY M AA R K S
REPAIR	R IY P EH R
REPAIRED	R IY P EH R DD

REPAIRING	R IY P EH R IX NG
REPLACED	R IY P L EY S TD
REPORT	R IY P AO R TD
REPORTED	R IY P AO R DX IX DD
REPORTING	R IY P AO R DX IX NG
REPORTS	R IY P AO R TS
RESET	R IY S EH TD
RESOLUTION	R EH Z AX L UW SH AX N
RESOLVED	R IX Z AA L V DD
RESOURCE	R IY S AO R S
RESOURCES	R IY S AO R S IX Z
REVIEW	R IY V Y UW
ROSS	R AA S
SACRAMENTO	S AE K R AX M EH N TD OW
SACRAMENTO'S	S AE K R AX M EH N TD OW Z
SAIL	S EY L
SAME	S EY M
SAMPLE	S AE M P L
SAMPLE'S	S AE M P L Z
SAN-DIEGO	S AE N D IY EY G OW
SAN-FRAN	S AE N F R AE N
SASSAFRAS	S AE S AX F R AE S
SASSAFRAS'S	S AE S AX F R AE S IX S
SATURDAY	S AE DX ER DX EY
SATURDAY'S	S AE DX ER DX EY Z
SAVE	S EY V
SCHENECTADY	S K AX N EH KD T IX DX IY
SCHENECTADY'S	S K AX N EH KD T IX DX IY Z
SCREEN	S K R IY N
SCREENS	S K R IY N Z
SEA	S IY
SEAWOLF	S IY W AO L F
SEAWOLF'S	S IY W AO L F S
SECOND	S EH K AX N DD
SECURITY	S AX K Y UH R IH DX IY
SENSOR	S EH N S ER
SENSORS	S EH N S ER Z
SEPTEMBER	S EH PD T EH M B ER
SET	S EH TD
SETTING	S EH DX IX NG
SETTINGS	S EH DX IX NG Z
SEVEN	S EH V AX N
SEVENTEEN	S EH V AX N T IY N
SEVENTEENTH	S EH V AX N T IY N TH
SEVENTH	S EH V AX N TH
SEVENTY	S EH V AX N DX IY
SHASTA	SH AE S T AX
SHASTA'S	SH AE S T AX Z

SHE	SH IY
SHE'S	SH IY Z
SHERMAN	SH ER M AX N
SHERMAN'S	SH ER M AX N Z
SHIP	SH IH PD
SHIP'S	SH IH P S
SHIPS	SH IH P S
SHIPS'S	SH IH P S
SHOW	SH OW
SHOWING	SH OW IX NG
SHOWN	SH OW N
SIBERIAN	S AY B IH R IY AX N
SIDNEY	S IH DD N IY
SIL	SIL
SINCE	S IH N S
SINGAPORE	S IH NG AX P AO R
SIX	S IH K S
SIXTEEN	S IH K S T IY N
SIXTEENTH	S IH K S T IY N TH
SIXTH	S IH K S TH
SIXTY	S IH K S T IY
SIZE	S AY Z
SLOW	S L OW
SLOWER	S L OW ER
SLOWEST	S L OW AX S TD
SLQ-32	EH S EH L K Y UW TH ER DX IY T UW
SMALL	S M AO L
SMALLER	S M AO L ER
SMALLEST	S M AO L IX S TD
SOHO	S OW HH OW
SOLOMON	S AA L AX M AX N
SONAR	S OW N AA R
SOON	S UW N
SOONER	S UW N ER
SOUTH	S AW TH
SOUTHERN	S AH DH ER N
SOVIET-UNION	S OW V IY EH TD Y UW N Y AX N
SPEED	S B IY DD
SPEEDS	S B IY D Z
SPS-40	EH S P IY EH S F AO R DX IY
SPS-48	EH S P IY EH S F AO R DX IY EY TD
SQQ-23	EH S K Y UW K Y UW T W EH N IY TH R IY
START	S D AA R TD
STARTED	S D AA R DX IX DD
STARTING	S D AA R DX IX NG
STATION	S D EY SH AX N
STATUS	S D AE DX AX S

STEAM	S D IY M
STEREOGRAPHIC	S D EH R IY IX G R AE F IX KD
STERETT	S D EH R IX TD
STERETT'S	S D EH R IX TS
STRAIT	S D R EY TD
SUB	S AH B
SUB'S	S AH B Z
SUBIC	S UW B IH K
SUBMARINE	S AH B M ER IY N
SUBMARINE'S	S AH B M ER IY N Z
SUBMARINES	S AH B M ER IY N Z
SUBMARINES'S	S AH B M ER IY N Z
SUBS	S AH B Z
SUBS'S	S AH B Z
SUFFICIENT	S AX F IH SH AX N TD
SUMMARIZE	S AH M ER AY Z
SUNDAY	S AH N D EY
SUNDAY'S	S AH N D EY Z
SUPPLIES	S AX P L AY Z
SUPPLY	S AX P L AY
SUPPOSED	S AX P OW Z DD
SURFACE	S ER F IX S
SUSTAINED	S AX S D EY N DD
SWITCH	S W IH CH
SWITCHES	S W IH CH IX Z
SWORDFISH	S AO R DD F IH SH
SWORDFISH'S	S AO R DD F IH SH IX Z
SYSTEM	S IH S T AX M
T-LAM	T IY L AE M
TACAN	T AE K IX N
TAIWAN	T AY W AA N
TAKE	T EY K
TAKEN	T EY K AX N
TASM	T AE S AX M
TEN	T EH N
TENTH	T EH N TH
TEST	T EH S TD
TEXAS	T EH K S IX S
TEXAS'S	T EH K S IX S IX Z
TFCC	T IY EH F S IY S IY
THAILAND	T AY L AE N DD
THAN	DH EH N
THAT	DH AE TD
THE	DH AX
THEIR	DH EH R
THEM	DH EH M
THERE	DH EH R
THESE	DH IY Z

THEY	DH EY
THIRD	TH ER DD
THIRTEEN	TH ER T IY N
THIRTEENTH	TH ER T IY N TH
THIRTIETH	TH ER DX IY AX TH
THIRTY	TH ER DX IY
THIS	DH IH S
THOSE	DH OW Z
THOUSAND	TH AW Z AX N DD
THREAT	TH R EH TD
THREATS	TH R EH TS
THREE	TH R IY
THURSDAY	TH ER Z D EY
THURSDAY'S	TH ER Z D EY Z
TICONDEROGA	T AY K AA N D AX R OW G AX
TICONDEROGA'S	T AY K AA N D AX R OW G AX Z
TIME	T AY M
TIMES	T AY M Z
TO	T UW
TODAY	T IX DX EY
TODAY'S	T IX DX EY Z
TOGGLE	T AA G L
TOGGLED	T AA G L DD
TOGGLING	T AA G L IX NG
TOKYO	T OW K IY OW
TOMORROW	T AX M AA R OW
TOMORROW'S	T AX M AA R OW Z
TONKIN	T AO NG K IX N
TONS	T AH N Z
TOTAL	T OW DX L
TOWNSVILLE	T AW N Z V IX L
TRACK	T R AE KD
TRACKS	T R AE K S
TRAINING	T R EY N IX NG
TRIPOLI	T R IH P AX L IY
TRIPOLI'S	T R IH P AX L IY Z
TRUE-VIEW	T R UW V Y UW
TUESDAY	T UW Z D EY
TUESDAY'S	T UW Z D EY Z
TURBINE	T ER B IX N
TURN	T ER N
TURNED	T ER N DD
TURNING	T ER N IX NG
TUSCALOOSA	T AX S K AX L UW S AX
TUSCALOOSA'S	T AX S K AX L UW S AX Z
TWELFTH	T W EH L F TH
TWELVE	T W EH L V
TWENTIETH	T W EH N IY IX TH

TWENTY	T W EH N IY
TWO	T UW
TYPE	T AY PD
TYPES	T AY P S
UNIT	Y UW N IH TD
UNITED-STATES	Y UW N AY DX IX DD S D EY TS
UNTIL	AX N T IX L
UOM	Y UW OW EH M
UPDATE	AH PD D EY TD
UPDATED	AH PD D EY DX IX DD
UPDATES	AH PD D EY TS
UPGRADE	AH PD G R EY DD
UPGRADED	AH PD G R EY DX IX DD
USA	Y UW EH S EY
USE	Y UW Z
USING	Y UW Z IX NG
USN	Y UW EH S EH N
VALUE	V AE L Y UW
VALUES	V AE L Y UW Z
VANCOUVER	V AE N K UW V ER
VANCOUVER'S	V AE N K UW V ER Z
VANDERGRIFT	V AE N D ER G R IH F TD
VANDERGRIFT'S	V AE N D ER G R IH F TS
VARIOUS	V EH R IY IX S
VESSEL	V EH S L
VESSEL'S	V EH S L Z
VESSELS	V EH S L Z
VESSELS'S	V EH S L Z
VIRGINIA	V ER JH IH N Y AX
VIRGINIA'S	V ER JH IH N Y AX Z
VISUAL	V IH SH UW L
WABASH	W AA B AE SH
WABASH'S	W AA B AE SH IX Z
WADSWORTH	W AA D Z W ER TH
WADSWORTH'S	W AA D Z W ER TH S
WAS	W AH Z
WASN'T	W AH Z AX N TD
WASP	W AA S PD
WASP'S	W AA S P S
WE	W IY
WEDNESDAY	W EH N Z D EY
WEDNESDAY'S	W EH N Z D EY Z
WEEK	W IY KD
WEEK'S	W IY K S
WEEKS	W IY K S
WELLINGTON	W EH L IH NG T AX N
WENT	W EH N TD
WERE	W ER

WEREN'T	W ER N TD
WEST	W EH S TD
WESTERN	W EH S T ER N
WESTPAC	W EH S TD P AE KD
WESTPAC'S	W EH S TD P AE K S
WHAT	W AH TD
WHAT'RE	W AH DX ER
WHAT'S	W AH TS
WHEN	W EH N
WHEN'LL	W EH N IX L
WHEN'S	W EH N Z
WHERE	W EH R
WHERE'S	W EH R Z
WHICH	W IH CH
WHIPPLE	W IH P L
WHIPPLE'S	W IH P L Z
WHO	HH UW
WHO'S	HH UW Z
WHOSE	HH UW Z
WHY	W AY
WHY'S	W AY Z
WICHITA	W IH CH IX T AA
WICHITA'S	W IH CH IX T AA Z
WILL	W IH L
WILLAMETTE	W IH L AX M EH TD
WILLAMETTE'S	W IH L AX M EH TS
WINAMAC	W IH N AX M AE KD
WINAMAC'S	W IH N AX M AE K S
WINDOW	W IH N D OW
WINDOWS	W IH N D OW Z
WITH	W IX TH
WITHIN	W IX TH IX N
WITHOUT	W IX DH AW TD
WON'T	W OW N TD
WORSE	W ER S
WORST	W ER S TD
WOULD	W UH DD
WOULDN'T	W UH DD AX N TD
YANKEE	Y AE NG K IY
YEAR	Y IH R
YEARS	Y IH R Z
YELLOW	Y EH L OW
YESTERDAY	Y EH S T ER DX EY
YESTERDAY'S	Y EH S T ER DX EY Z
YET	Y EH TD
YORKTOWN	Y AO R KD T AW N
YORKTOWN'S	Y AO R KD T AW N Z
ZERO	Z IY R OW

`ZULU                    Z UW L UW`

II.2. The Grammar

The resource management grammar is defined by a set of 900 templates. To conserve space, we list the first 10 templates to provide a flavor of the task (italicized words bracketed by angle brackets are non-terminals):

<what-is> *<optthe>* *<shipname's>* *<gross~ave>* displacement in *<long~metric>* tons

is *<optthe>* *<shipname's>* *<earliest>* *<casrep>* rated worse than *<optthe>* hers

<list> *<optthe>* *<threats>*

<list> *<optthe>* *<shipname's>* *<casreps>* from the last *<digit>* months

<show-list> *<optthe>* *<shipname's>* home port

<draw-show> *<optthe>* *<shipname's>* last *<digit>* *<sensor>* *<posits>*

is *<optthe>* *<shipname's>* remaining fuel insufficient to arrive in port at *<optthe>* *<current>* speed

<list> *<optthe>* *<shipname's>* *<gross~ave>* displacement and capabilities

<draw-show> *<optthe>* *<shipname's>* track in *<bright-dim>* *<color>* with *<optthe>* *<shipname's>* in *<bright-dim>* *<color>*

is *<optthe>* *<shipname's>* fuel capacity *<greater-than>* *<optthe>* *<shipname's>*

II.3. Training and Test Speakers

Table II-1 enumerates all 120 speakers released by TI to CMU. Among these speakers, the *80 training speakers* and the *first 25 evaluation speakers* were used to train SPHINX. The *10 March-87 evaluation speakers* and the *6 Oct-87 evaluation speakers* were used to test SPHINX. Since one speaker is overlapped between the two evaluation sets, there are actually 15 test speakers. The first four characters of a speaker ID identify the speaker, the next digit encodes the dialect of the speaker, and the last letter indicates male or female, where available.

Each of the speakers spoke 40 sentences. For the training speakers, all 40 sentences were used to train the HMMs. For the testing speakers, 10 of the sentences were designated by DARPA as evaluation sentences, and were used only for final tests. The remaining 30 sentences were used to tune the parameters of SPHINX.

80 training speakers							
adg04f	ahh05m	aks01f	apv03m	bar07m	bas04f	bef03m	bjk02m
bma04m	bmh05f	bns04m	bom07m	bwm03m	bwp04m	cal03m	cef03m
ceg08f	cft04f	cke03f	cmb05m	cmr02f	crc05m	csh03m	cth07m
cyl02f	das05m	daw18m	dhs03m	djh03f	dlb02m	dlh03m	dlr07m
dlr17m	dms04f	dmt02m	drd06m	dsc06m	dtb03m	eeh04f	ejs08m
ers07m	etb01f	fwk04m	gjd04f	gmd05f	gxp04m	hbs07m	hes05f
hpg03m	jcs05f	jem01f	jma02m	jpg05m	jrk06m	jws04m	jxm05f
kes06m	kkh05f	lih05m	ljc04m	mah05f	mcc05m	mdm04m	mgk02m
mju06f	mmh02f	pgh01m	pgl02m	rcg01m	rgm04m	rkm01m	rtk03m
rws01m	sdc05f	tju06m	tlb03m	tpf01m	utb05f	vlo05m	wem05m

First 25 evaluation speakers								
ajp06	bgt05	bpm05	cae06	chh07	cpm01	dtd05	ejl06	esd06
esj06	grl01	hjb03	hxs06	jlm04	jln08	jmd02	jsa05	lag06
rav05	rdd01	rjm12	sds06	tab07	tdp05	wbt01		

March-87 evaluation speakers				
awf05	bcg18	bth07	ctt05	dab01
dlc03	gwt02	jfc06	jfr07	sah01

Oct-87 evaluation speakers					
ctm02	dpk01	gwt02	ljd03	lmk05	sjk06

Table II-1: The list of all 120 speakers released by TI.

Appendix III
Examples of SPHINX Recognition

In this appendix, we enumerate the results of SPHINX on the 150 test sentences, using the best SPHINX configuration described in Table 6-11. For each sentence, we show the correct sentence, as well as the recognized sentence using the bigram grammar, the word-pair grammar, and no grammar. Each word error is italicized, and insertions are designated with **

Correct: what's the mercury's average cruising speed
Bigram: what's the mercury's average cruising speed
Word-pair: what's the mercury's average cruising speed
None: what's the mercury's average cruising speed

Correct: what is the eta at her destination of fanning
Bigram: what is the eta at her destination of fanning
Word-pair: what is the eta at her destination of fanning
None: what is *it* eta at her destination of fanning

Correct: how soon can esteem chop to atlantic fleet
Bigram: how soon can esteem chop to atlantic fleet
Word-pair: how soon can esteem chop to atlantic fleet
None: how *sea again* esteem chop to atlantic fleet

Correct: are there no ships that are in the mozambique channel
Bigram: ** *find any* ships that are in the mozambique channel
Word-pair: *find the nine* ships that are in the mozambique channel
None: *by end as* ships that *their and* the mozambique channel

Correct: draw a chart of ross sea
Bigram: draw ** chart of ross sea
Word-pair: draw ** chart of ross sea
None: draw ** chart *overall* ** sea

Correct: what was peoria's location and asuw area mission code july one
Bigram: what was peoria's location and asuw area mission code july one
Word-pair: what was peoria's location and asuw area mission code july one
None: what was peoria's location and asuw area mission *ten* july one

Correct: is rathburne located in wellington or aberdeen
Bigram: is rathburne located in wellington or aberdeen
Word-pair: is rathburne located in wellington or aberdeen
None: is rathburne located in wellington *more* aberdeen

Correct: what frigates in bering sea have both lamps and sps-48
Bigram: what frigates in bering sea have both lamps and sps-48
Word-pair: what frigates in bering sea have both lamps and sps-48
None: what frigates *the* bering sea have *bad* lamps *in* sps-48

Correct:	what frigates in bering sea have both lamps and sps-48
Bigram:	what frigates in bering sea have both lamps and sps-48
Word-pair:	what frigates in bering sea have both lamps and sps-48
None:	what frigates *the* bering sea have *bad* lamps *in* sps-48
Correct:	display the tracks and speeds of ships that are in solomon sea
Bigram:	display the tracks and speeds of ships that are in solomon sea
Word-pair:	display the tracks and speeds of ships that are in solomon sea
None:	display the *track since* speeds *the* ships that are *dim* solomon sea
Correct:	display a chart centered around jarrett using stereographic projection
Bigram:	display a chart centered around jarrett using stereographic projection
Word-pair:	display ** chart centered around jarrett using stereographic projection
None:	display ** chart *set add* around jarrett using stereographic projection
Correct:	what link-11 cruisers are in sea of japan
Bigram:	what link-11 cruisers are in sea of japan
Word-pair:	what *ntds and* cruisers are in sea of japan
None:	what *the yankee adding* cruisers are in sea *on* japan
Correct:	display the tracks of any cruisers in eastpac
Bigram:	display the tracks of any cruisers in eastpac
Word-pair:	display the tracks of any cruisers in eastpac
None:	display the tracks *would* any cruisers in eastpac
Correct:	will firebush be at miami tomorrow
Bigram:	will firebush be at miami tomorrow
Word-pair:	will firebush be at miami tomorrow
None:	will firebush be *yet* miami tomorrow
Correct:	what is mishawaka's percent fuel
Bigram:	what is mishawaka's percent fuel
Word-pair:	what is mishawaka's percent fuel
None:	what *these* mishawaka's *to set* fuel
Correct:	did mob mission area of the copeland ever go to m4 in nineteen eighty one
Bigram:	did mob mission area of the copeland ever go to m4 in nineteen eighty one
Word-pair:	*give* mob mission area of ** copeland ever go to m4 in nineteen eighty one
None:	did mob mission *carry echo* the *code went* ever go to m4 in nineteen *east* one
Correct:	are there more than four sps-40 capable frigates in port now
Bigram:	are there more than four sps-40 capable frigates in port now
Word-pair:	are there more than four sps-40 capable frigates in port now
None:	*by* their more *bad full aren't* sps-40 capable frigates *than* port now
Correct:	total the ships that will arrive in diego-garcia by next month
Bigram:	total the ships that will arrive in diego-garcia by next month
Word-pair:	total the ships that will arrive in diego-garcia by next month
None:	total *a* ships's *pac* will *arriving* ** diego-garcia by next *on*
Correct:	redisplay overlay soho turning on *echo*
Bigram:	redisplay overlay soho turning on *april*
Word-pair:	redisplay overlay soho turning on *pluck now*
None:	redisplay *overlays* soho turning *lon pac now*

Correct: is mcclusky's destination the same as ramsey's
Bigram: is mcclusky's destination the same as ramsey's
Word-pair: is mcclusky's destination the same as ramsey's
None: is mcclusky's destination ** *esteem has* ramsey's

Correct: show on data screen arkansas's track since four october
Bigram: show *lon* data screen arkansas's track since four october
Word-pair: show *lon* data screen arkansas's track since four october
None: show *lon beam* screen arkansas's track since four october

Correct: is jason's maximum sustained speed slower than jupiter's
Bigram: is jason's maximum sustained speed slower than jupiter's
Word-pair: is jason's maximum sustained speed slower than jupiter's
None: is jason's maximum sustained *speeds* slower than jupiter's

Correct: which submarines in bismark sea have tacan
Bigram: which submarines in bismark sea have tacan
Word-pair: which submarines in bismark sea have tacan
None: which submarines *an* bismark sea have *if tonkin*

Correct: what is the name and c-code of the carrier in siberian sea
Bigram: what is the name and c-code of the carrier in siberian sea
Word-pair: what is the name and c-code of the carrier in siberian sea
None: what is the name and c-code of the carrier in siberian sea

Correct: do any ships that are in bass strait have an m5 miw m-rating
Bigram: do any ships that are in bass strait have an m5 miw m-rating
Word-pair: do any ships that are in bass strait have an m5 miw m-rating
None: *to rated* ships that ** *during by* strait have *the the* m5 *of* miw *won't* m-rating

Correct: set the color of hooked track to bright red
Bigram: set ** color of hooked track to bright red
Word-pair: set the color of hooked track to bright red
None: set the color ** *cook chart* to *brooke* red

Correct: is constant's last location closer than denver's to pac alert
Bigram: is constant's last location closer than denver's to pac alert
Word-pair: is constant's last location closer than denver's to pac alert
None: is constant's last *the* location closer than denver's to *current* alert

Correct: is copeland farther from sidney than the davidson
Bigram: is copeland farther from sidney than the davidson
Word-pair: is copeland farther from sidney than the davidson
None: is *copeland's earlier* from sidney than the davidson

Correct: how far is the meteor from the midgett
Bigram: how far is the meteor from ** midgett
Word-pair: how far is ** meteor from ** midgett
None: how far is ** meteor from ** midgett

Correct: redefine area pac
Bigram: redefine area pac
Word-pair: redefine area pac
None: *redefined* area pac

Correct:	how close is ** gulf of california to davidson
Bigram:	how close is *the* gulf of california to davidson
Word-pair:	how close is gulf of california to davidson
None:	how close *of* is gulf *if* california to davidson
Correct:	when did seawolf degrade from her previous equipment c-rating
Bigram:	when did seawolf degrade from her previous equipment c-rating
Word-pair:	when did seawolf degrade from her previous equipment c-rating
None:	when did seawolf degrade from her previous equipment c-rating
Correct:	what frigate in north atlantic ocean has the slowest current speed
Bigram:	what frigate in north atlantic ocean has the slowest current speed
Word-pair:	what frigate in north atlantic ocean has the slowest current speed
None:	what frigate *the* north *lat* ocean has ** slowest *kirk* speed
Correct:	how many submarines were in port-victoria on the twentieth of march
Bigram:	how many submarines were in port-victoria on the twentieth of march
Word-pair:	how many submarines were in port-victoria *of homer* twentieth of march
None:	how many submarines *weren't* ** port-victoria ** *homer* twentieth of march
Correct:	display a new chart projection using mercator
Bigram:	display a new chart projection using mercator
Word-pair:	display a new chart projection using mercator
None:	display ** new chart projection using mercator
Correct:	show grill
Bigram:	show grill
Word-pair:	show grill
None:	show grill
Correct:	get the cruiser's locations for april
Bigram:	get the cruiser's locations for april
Word-pair:	get the cruiser's locations for april
None:	get ** cruisers locations for april
Correct:	what is the average training rating code for usn ships that are in arctic ocean
Bigram:	what is the average training rating code for usn ships that are in arctic ocean
Word-pair:	what is the average training rating *codes* for usn ships ** are in arctic ocean
None:	what *using* ** average training *green* code for usn ships ** *sterett* ** arctic ocean
Correct:	reset the switches to defaults
Bigram:	reset ** switches to defaults
Word-pair:	reset ** switches to defaults
None:	reset ** switches to defaults
Correct:	edit location data for track a42128
Bigram:	edit location data for track a42128
Word-pair:	edit location data for track a42128
None:	edit location data for track a42128
Correct:	do any vessels that are in gulf of tonkin have asw mission area of m4
Bigram:	do any vessels that are in gulf of tonkin have asw mission area of m4
Word-pair:	do any vessels that are in gulf of tonkin have asw mission area of m4
None:	*to* any vessel's ** ** *bering* gulf *both* tonkin have asw mission area *both* m4

Correct: what's glacier's maximum draft
Bigram: what's glacier's maximum draft
Word-pair: what's glacier's maximum draft
None: what's glacier's maximum draft

Correct: where was the brooke on january sixteen
Bigram: where was the brooke on january sixteen
Word-pair: where was the brooke *were in* january sixteen
None: *where's* ** the brooke *which years* sixteen

Correct: clear data screen
Bigram: clear data screen
Word-pair: clear data screen
None: *cleared did* screen

Correct: how many vessels are in indian ocean
Bigram: how many vessels are in indian ocean
Word-pair: how many vessels are in indian ocean
None: how many vessels are ** indian ocean

Correct: what is the midway's fuel level
Bigram: what is the midway's fuel level
Word-pair: what is the midway's fuel level
None: what *list* the midway's fuel *levels*

Correct: show the conquest's position seventeen august of eighty six
Bigram: show the conquest's position seventeen august *ten* eighty six
Word-pair: show the conquest's position seventeen august *ten* eighty six
None: show the *conquest* position seventeen ** *more estimated* six

Correct: why did queenfish change equipment readiness twenty three may
Bigram: why did queenfish change equipment readiness twenty three may
Word-pair: why did queenfish change equipment readiness twenty three may
None: why *eight* queenfish change equipment readiness twenty three may

Correct: show percent fuel aboard mercury
Bigram: show percent fuel aboard mercury
Word-pair: show percent fuel aboard mercury
None: show percent fuel aboard *were three*

Correct: find crovls and tracks for tfcc frigates in north pacific ocean
Bigram: find crovls and tracks for tfcc frigates in north pacific ocean
Word-pair: find crovls and tracks for tfcc frigates in north pacific ocean
None: find crovls end tracks for tfcc frigates ** north pacific ocean

Correct: toggle sail and save switches
Bigram: toggle sail and save switches
Word-pair: toggle sail and save switches
None: toggle sail *an sea* switches

Correct: where was frederick's destination november fourteenth
Bigram: where was frederick's destination november fourteenth
Word-pair: where was frederick's destination november fourteenth
None: where was frederick's destination november fourteenth

Correct:	has swordfish reported any training problems
Bigram:	has swordfish reported any training problems
Word-pair:	has swordfish reported any training problems
None:	has swordfish reported any training problems
Correct:	is tripoli's fuel capacity larger than tuscaloosa's
Bigram:	is tripoli's fuel capacity larger than tuscaloosa's
Word-pair:	is tripoli's fuel capacity larger than tuscaloosa's
None:	is tripoli's fuel capacity larger than tuscaloosa's
Correct:	how many vessels were deployed since thirty one october
Bigram:	how many vessels were deployed since thirty one october
Word-pair:	how many vessels were deployed since thirty one october
None:	how many vessels were deployed since thirty one october
Correct:	set switches to their defaults
Bigram:	set switches to their defaults
Word-pair:	set switches to their defaults
None:	set switches to their defaults
Correct:	display virginia's displacement in metric tons
Bigram:	display virginia's displacement in metric tons
Word-pair:	display virginia's displacement in metric tons
None:	display virginia's displacement in metric tons
Correct:	show posits of frigates that are in westpac
Bigram:	show posits of frigates that are in westpac
Word-pair:	show posits of frigates that are in westpac
None:	show posits of frigates that are *again* westpac
Correct:	show supplies readiness of ironwood august one
Bigram:	show supplies readiness of ironwood august one
Word-pair:	show supplies readiness of ironwood august one
None:	show supplies readiness *of* of *ironwood's* august one
Correct:	how early can fox be there
Bigram:	how early can fox be there
Word-pair:	how early can fox be there
None:	how *all* early can fox *beam* there
Correct:	find full position data for all tracks
Bigram:	find full position data for all tracks
Word-pair:	find full position data for all tracks
None:	find full position data four all tracks
Correct:	what is midgett's percent fuel
Bigram:	what is midgett's percent fuel
Word-pair:	what is midgett's percent fuel
None:	what *he's* midgett's percent fuel
Correct:	which ships in manchester have a supplies readiness rating of c5
Bigram:	which ships in manchester have *an* supplies readiness rating of c5
Word-pair:	which ships *due* in manchester have *an* supplies readiness rating of c5
None:	which ships's *dim* manchester have ** supplies readiness rating of c5

Correct: what casrep did firebush have on twenty seven may
Bigram: what casrep did firebush have on twenty seven may
Word-pair: what casrep did firebush have on twenty seven may
None: what casrep did firebush have *one* twenty seven may

Correct: how many ships are not ntds capable
Bigram: how many ships are not ntds capable
Word-pair: how many ships are not ntds capable
None: *home* many ships are knot ntds capable

Correct: clear display
Bigram: clear display
Word-pair: clear display
None: clear display

Correct: is pigeon's test depth greater than pluck's
Bigram: is pigeon's test depth greater than pluck's
Word-pair: is pigeon's test depth greater than pluck's
None: is pigeon's test depth greater than pluck's

Correct: is citrus more than eighty kilometers from cleveland
Bigram: is citrus more than eighty kilometers from cleveland
Word-pair: is citrus more than eighty kilometers from cleveland
None: is citrus more than eighty kilometers from cleveland

Correct: what is the distance from the mishawaka to the monticello
Bigram: what is the distance from ** mishawaka to the monticello
Word-pair: what is the distance from ** mishawaka to the monticello
None: *went* is the distance from ** mishawaka to the monticello

Correct: what is the name and the various ratings of the frigate in west bering sea
Bigram: what is the name and the various ratings of the frigate in west bering sea
Word-pair: what is the name and the various ratings of the frigate in west bering sea
None: *two* what is ** *sustained* and the various ratings of the frigate in west bering sea

Correct: which subs that are c3 are in korean bay
Bigram: which subs that are c3 are in korean bay
Word-pair: which subs that are c3 are in korean bay
None: which sub's ** *sooner* c3 are *reporting* ** bay

Correct: get the ships and their fleet identifications
Bigram: get the ships *in* their fleet identifications
Word-pair: get the ships *in* their fleet identifications
None: get the ships *do* their fleet identifications

Correct: show the various fleet identifications for frigates
Bigram: show the various fleet identifications for frigates
Word-pair: show the various fleet identifications for frigates
None: show the various fleet identifications for frigates

Correct: on what day ** could dubuque arrive in port at his maximum sustained speed
Bigram: ** what *would it take* dubuque arrive in port at his maximum sustained speed
Word-pair: on what day *can* dubuque arrive in port at his maximum sustained speed
None: *her would date the* dubuque *arriving* ** port ** *hers* maximum sustained speed

Correct:	what is manhattan's fuel capacity
Bigram:	what is manhattan's fuel capacity
Word-pair:	what is manhattan's fuel capacity
None:	*would* is manhattan's fuel capacity
Correct:	what is the asw average rating code for ships in formosa strait
Bigram:	what is the asw average rating code for ships in formosa strait
Word-pair:	what is the asw average rating code for ships in formosa strait
None:	what is ** asw average rating code for *ship show* formosa strait
Correct:	weren't more than ninety c1 ships in pacific fleet today
Bigram:	*were* more than ninety c1 ships in pacific fleet today
Word-pair:	*were* more than ninety c1 ships in pacific fleet today
None:	*were* more *the* ninety *she one* ships *of* pacific fleet *to at*
Correct:	how many kilometers is anchorage from new-york
Bigram:	how many kilometers is anchorage *of* new-york
Word-pair:	how many kilometers is anchorage *to* new-york
None:	how many kilometers *does* anchorage ** new-york
Correct:	draw the tracks of all subs that are in gulf of tonkin
Bigram:	draw ** tracks *for* all subs that are in gulf of tonkin
Word-pair:	draw ** tracks of all subs that are in gulf of tonkin
None:	draw ** tracks ** *will* subs's ** ** *centering* gulf ** tonkin
Correct:	show queenfish's location on twenty two february and its various capabilities
Bigram:	show queenfish's location on twenty two february and its various capabilities
Word-pair:	show queenfish's location on twenty two february and its various capabilities
None:	show *queenfish* location on twenty two *from weren't minutes hers* capabilities
Correct:	who has the least fuel remaining
Bigram:	who has ** least fuel remaining
Word-pair:	who has ** least fuel remaining
None:	who has ** least fuel remaining
Correct:	give any cruisers that were c2 on eight august
Bigram:	give any cruisers that were c2 on eight august
Word-pair:	give any cruisers that were c2 on eight august
None:	give any cruisers *the* were *she two lon* eight august
Correct:	give c5 ships in pacific fleet
Bigram:	give c5 ships in pacific fleet
Word-pair:	give c5 ships in pacific fleet
None:	give c5 ships *an* pacific fleet
Correct:	what is the number of vessels that are in ross sea without slq-32
Bigram:	what is the number of vessels that are in ross sea without slq-32
Word-pair:	what is the number of vessels that are in ross sea *with pollack the* slq-32
None:	what is ** number ** vessels ** *letter* ** ross sea *with pollack* slq-32
Correct:	turn groups on and redraw the current area
Bigram:	turn groups on and redraw the current area
Word-pair:	turn groups on and redraw the current area
None:	turn groups *long than* redraw the current area

Correct:	which vessels in korean bay have a supplies readiness that is c3
Bigram:	which vessels in korean bay have *any* supplies readiness ** *of* c3
Word-pair:	which vessels in korean bay have *any* supplies readiness *the* ** c3
None:	which vessels in korean bay have *he* supplies readiness *the* ** c3
Correct:	set the sail parameter to off
Bigram:	set ** sail parameter *turned* off
Word-pair:	set ** sail parameter *turned* off
None:	*show personnel* ** parameter to *of* off
Correct:	which link-11 capable carriers have an equipment resource rating of more than c4
Bigram:	which link-11 capable carriers have an equipment resource rating of more than c4
Word-pair:	which link-11 capable carriers have an equipment resource rating of more than c4
None:	which link-11 capable carriers *haven't* ** equipment resource rating of *in were* than c4
Correct:	set switches to defaults
Bigram:	set switches to defaults
Word-pair:	set switches to defaults
None:	*show* switches to defaults
Correct:	what if apalachicola's propulsion type was steam turbine instead of gas
Bigram:	what if apalachicola's propulsion type was steam turbine instead of gas
Word-pair:	what if apalachicola's propulsion type was steam turbine instead of gas
None:	*one give* apalachicola's propulsion type ** steam turbine instead of *get bass*
Correct:	list vessels that were deployed on the first of september
Bigram:	list vessels that *weren't* deployed on ** first of september
Word-pair:	list vessels *and weren't* deployed on ** first of september
None:	list vessels *an* were deployed *conifer's* ** ** of september
Correct:	find frigates in honolulu
Bigram:	find frigates in honolulu
Word-pair:	find frigates in honolulu
None:	find frigates *end* honolulu
Correct:	show the same chart with nova
Bigram:	show the same chart with nova
Word-pair:	show the same chart with nova
None:	show ** *sustained* chart with nova
Correct:	show the names of any submarines in yellow sea on twenty eight october
Bigram:	show the names of any submarines in yellow sea on twenty ** october
Word-pair:	show the names of any submarines in yellow sea on twenty ** october
None:	show the names *have* any submarine's *beam* yellow sea on twenty *eighty* october
Correct:	why was mercury's miw m-code changed on april twenty two
Bigram:	why was mercury's miw m-code changed on april twenty two
Word-pair:	why was mercury's miw m-code changed on april twenty two
None:	why was *where two recent* miw m-code *change* on april twenty two
Correct:	which ships in philippine sea are link-11 capable
Bigram:	which ships in philippine sea *aren't* link-11 capable
Word-pair:	which ships in philippine sea *aren't* link-11 capable
None:	which ships in philippine ** *c5* link-11 capable

Correct:	is there a problem with personnel for the camden
Bigram:	is there a problem with personnel *from* the camden
Word-pair:	is there a problem *is* personnel for *all lat is* camden
None:	is *bering* ** *problems* ** personnel *far lat* camden
Correct:	are there any cruisers longer than nineteen hundred meters that are in siberian sea
Bigram:	** *find* any cruisers longer than nineteen hundred meters that are in siberian sea
Word-pair:	** *find* any cruisers longer than nineteen hundred meters that are in siberian sea
None:	*gone* there *many* cruisers *lon in* than nineteen hundred *me is* that *areas* ** siberian sea
Correct:	increase letter size to maximum value and redraw
Bigram:	increase letter size to maximum value and redraw
Word-pair:	increase letter size to maximum value and redraw
None:	increase *lat* size ** maximum value *in* redraw
Correct:	show downes's radar latitudes and longitudes using nova
Bigram:	show downes's radar latitudes and longitudes using nova
Word-pair:	show downes's radar latitudes and longitudes using nova
None:	show *downes* radar latitudes *in* longitudes using nova
Correct:	find me the mission area ratings for arkansas
Bigram:	find me the mission area ratings for arkansas
Word-pair:	find me the mission area ratings for arkansas
None:	find *these than* mission area ratings *five* arkansas
Correct:	turn areas off and redraw current area
Bigram:	turn areas off and redraw current area
Word-pair:	turn areas off and redraw current area
None:	turn areas *half in* redraw current area
Correct:	what is the mob m-code for sample
Bigram:	what is the mob m-code for sample
Word-pair:	what is the mob m-code for sample
None:	what is *be* mob *ten code* for sample
Correct:	how soon does fresno arrive in townsville
Bigram:	how soon does fresno arrive in townsville
Word-pair:	how soon does fresno arrive in townsville
None:	** *hasn't is* fresno arrive in townsville
Correct:	were there more than fifteen pacific fleet vessels employed in nineteen eighty three
Bigram:	were there more than fifteen pacific fleet vessels employed in nineteen eighty three
Word-pair:	were there more than fifteen pacific fleet vessels *been* employed in nineteen eighty three *for fifth*
None:	were there more than fifteen pacific fleet vessels *deployed* ** nineteen eighty three *off*
Correct:	is the wasp's last location closer than wichita's to osgp
Bigram:	is the wasp's last location closer than wichita's to osgp
Word-pair:	is the wasp's last location closer than wichita's to osgp
None:	is ** *last* last location closer than wichita's to *thailand list cep*
Correct:	has ** horne's miw mission area gone to m3 before twenty two august
Bigram:	has *mars on* miw mission area gone to m3 before twenty two august
Word-pair:	has horne's miw mission area gone to m3 before twenty two august
None:	*it* has horne's miw mission area *going* two m3 before twenty two august

Correct:	display a chart of bering sea with the time window from twenty four hundred to eighteen hundred hours zulu
Bigram:	display a chart of bering sea with ** time window from twenty four hundred to eighteen hundred hours zulu
Word-pair:	display ** chart of bering sea with ** time window from twenty four hundred to eighteen hundred hours zulu
None:	display ** chart ** bering sea with ** time *in that* from twenty four *had eight it* eighteen hundred hours *beam*

Correct:	how many in west philippine sea have more than half their fuel left
Bigram:	how many in west philippine sea have more than half their fuel left
Word-pair:	how many in west philippine sea have more than half their fuel left
None:	how *any* in west philippine sea have more than half *the* fuel left

Correct:	how close is seawolf's last location to fifty two degrees north eight degrees east
Bigram:	how close is seawolf's last location to fifty ** degrees north eight degrees east
Word-pair:	how close is seawolf's last location to fifty ** degrees north eight degrees east
None:	how close ** seawolf's last location two fifty ** degrees north *be decrease* east

Correct:	get the destinations and arrival hour at destination for all subs
Bigram:	get the destinations and arrival hour at destination for all subs
Word-pair:	get ** destinations and arrival hour at destination for all subs
None:	get the destinations ** arrival hour *bad* destination for all subs

Correct:	get all usn ships that are in coral sea
Bigram:	get all usn ships that are in coral sea
Word-pair:	get all usn ships that are in coral sea
None:	*give* all usn ships that ** *bering* coral sea

Correct:	turn off cep switch
Bigram:	turn off cep switch
Word-pair:	turn off cep switch
None:	turn *above* cep switch

Correct:	is the economic speed of apalachicola less than that of the brunswick
Bigram:	** *whose* economic speed of apalachicola less than that of the brunswick
Word-pair:	** *whose* economic speed of apalachicola *a list the lat* of ** brunswick
None:	*his* ** economic speed *do* of apalachicola *list* ** ** ** the *level* brunswick

Correct:	define area alerts for gulf of california
Bigram:	define area alerts for gulf of california
Word-pair:	define area alerts for gulf of california
None:	define area alerts for gulf of california

Correct:	clear display window
Bigram:	clear display window
Word-pair:	clear display window
None:	clear display window

Correct:	show me horne's track in dim orange with reeves's in bright green
Bigram:	show me horne's track in dim orange with *reeves* in bright green
Word-pair:	show me horne's track in dim orange with *reeves* in bright green
None:	show me horne's track *indian* ** orange with *reeves* in bright green

Correct:	get resource area ratings for enterprise
Bigram:	get resource area ratings for enterprise
Word-pair:	get resource area ratings for enterprise
None:	*did* resource area ratings for enterprise

Correct:	is anybody in westpac ntds capable
Bigram:	is anybody in westpac ntds capable
Word-pair:	is anybody in westpac ntds capable
None:	is anybody in *less pac* ntds capable
Correct:	list monday's and tuesday's casualty reports for frigates in bass strait
Bigram:	list monday's and tuesday's casualty reports for frigates in bass strait
Word-pair:	list monday's and tuesday's casualty reports for frigates in bass strait
None:	list monday's and tuesday's casualty reports *far* frigates *can* bass strait
Correct:	how many cruisers that are sqq-23 capable are there at bombay
Bigram:	how many cruisers that *aren't* sqq-23 capable are there at bombay
Word-pair:	how many cruisers that *the* sqq-23 capable are there at bombay
None:	how many cruisers ** *sterett* sqq-23 capable *by* there *have* bombay
Correct:	which of the cruisers that are in korean bay have sps-48
Bigram:	which of the cruisers that are in korean bay have sps-48
Word-pair:	which of the cruisers that are in korean bay have sps-48
None:	which of *a* cruisers that are in korean bay have sps-48
Correct:	does the campbell have four open cat-3 problems
Bigram:	*was* the campbell have four open cat-3 problems
Word-pair:	does the campbell have four open cat-3 problems
None:	does the campbell have four open cat-3 problems
Correct:	has jason been downgraded yet
Bigram:	has jason been downgraded yet
Word-pair:	has jason been downgraded yet
None:	has jason *in* downgraded yet
Correct:	is there a gulf of thailand ship rated m5 on miw
Bigram:	is there a gulf of thailand ship rated m5 on miw
Word-pair:	is there a gulf of thailand ship rated m5 on miw
None:	is their ** gulf of thailand ship rated *turning thailand lon* miw
Correct:	what's hawkbill's fleet identification
Bigram:	what's hawkbill's fleet identification
Word-pair:	what's hawkbill's fleet identification
None:	what's hawkbill's fleet identification
Correct:	show on data screen ranger's track since october thirteenth
Bigram:	show *lon* data screen ranger's track since october thirteenth
Word-pair:	show *lon* data screen ranger's track since october thirteenth
None:	show *one dated* screen *queenfish* track since october *if* thirteenth
Correct:	what is the aaw rating of the virginia
Bigram:	what is the aaw rating of the virginia
Word-pair:	what is *fiji in* aaw rating of the virginia
None:	what is *eta go beam* rating *both* the virginia
Correct:	list all the alerts
Bigram:	list all ** alerts
Word-pair:	list all ** alerts
None:	list all *me* alerts

Correct:	get names and latitudes for sps-40 carriers in arabian sea twenty seven november
Bigram:	get names and latitudes for sps-40 carriers in arabian sea twenty seven november
Word-pair:	get names and latitudes for sps-40 carriers in arabian sea twenty seven november
None:	get names and latitudes for sps-40 carriers in arabian sea twenty seven november
Correct:	give vessels in indian ocean and their destinations
Bigram:	give vessels in indian ocean and their destinations
Word-pair:	give vessels in indian ocean and their destinations
None:	give vessel's in *only* ocean end their destinations
Correct:	where's pollack now
Bigram:	where's pollack now
Word-pair:	where's pollack now
None:	where's pollack now
Correct:	toggle the unit of measure parameter
Bigram:	toggle ** unit of measure parameter
Word-pair:	toggle ** unit of measure parameter
None:	toggle ** unit ** measure parameter
Correct:	are any ships in bismark sea below ninety percent of their fuel capacity
Bigram:	are any ships in bismark sea below *nine* percent of their fuel capacity
Word-pair:	are any ships in bismark sea below *nine* percent of their fuel capacity
None:	*where me* ships in bismark sea below *mind hers letter* their fuel capacity
Correct:	increase letter size to the max value and redraw
Bigram:	increase letter size to the max value and redraw
Word-pair:	increase letter size to ** max value and redraw
None:	increase letter size ** ** max value ** redraw *off*
Correct:	what's the cleveland's current readiness
Bigram:	what's the cleveland's current readiness
Word-pair:	what's the cleveland's current readiness
None:	what's the *cleveland start* readiness
Correct:	get latitudes and longitudes and names of ships in the arabian sea
Bigram:	*give* latitudes and longitudes and names of ships in ** arabian sea
Word-pair:	*give* latitudes and longitudes *of* names of ships in ** arabian sea
None:	*give* latitudes *in* longitudes ** names *an* ships *near* ** arabian sea
Correct:	find missions edited today
Bigram:	find missions edited today
Word-pair:	find missions edited today
None:	find missions edited today
Correct:	get the various capabilities for gas turbine ships in the gulf of tonkin
Bigram:	get the various capabilities for gas turbine ships in ** gulf of tonkin
Word-pair:	get the various capabilities for gas turbine ships in ** gulf of tonkin
None:	*give do* various capabilities for gas turbine ships *an* ** gulf ** tonkin
Correct:	was lockwood's location on sunday in sea of japan
Bigram:	was lockwood's location on sunday in sea of japan
Word-pair:	was lockwood's location on sunday in sea *in* japan
None:	was lockwood's location on sunday in sea ** japan

Correct:	get c2 ships that are in diego-garcia
Bigram:	get c2 ships that are in diego-garcia
Word-pair:	get c2 ships that are in diego-garcia
None:	get c2 ships ** ** *centering* diego-garcia
Correct:	review alerts within the last ten hours for the ships that are in gulf of alaska
Bigram:	review alerts within the last ten hours for the ships that are in gulf of alaska
Word-pair:	review alerts within the last ten hours for the ships that are in gulf of alaska
None:	*meteor no arctic than* the last ten hours *from* the ships that are in gulf of alaska
Correct:	what speed is eisenhower going
Bigram:	what speed is eisenhower going
Word-pair:	what speed is eisenhower going
None:	what speed is eisenhower going
Correct:	define an alert for the formosa strait
Bigram:	define an alert for the formosa strait
Word-pair:	define an alert for the formosa strait
None:	*defining all aren't* for the formosa strait
Correct:	show the same chart with time started at nineteen hundred zulu
Bigram:	show the same chart with time started at nineteen hundred zulu
Word-pair:	show the same chart with time started at nineteen hundred zulu
None:	show the same chart *weren't* time started *have* nineteen *end its england*
Correct:	is eisenhower's beam smaller than mississippi's
Bigram:	is *the* eisenhower's beam smaller than mississippi's
Word-pair:	is eisenhower's beam smaller than mississippi's
None:	is eisenhower's beam *small* than mississippi's
Correct:	show locations for subs in eastpac that went to c1 on eleven january
Bigram:	show locations for subs in eastpac that went to c1 on eleven january
Word-pair:	show locations for subs in eastpac that went to c1 on eleven january
None:	show locations for subs *and* eastpac that *one* to c1 on eleven january
Correct:	give current equipment readiness of the hector
Bigram:	give current equipment readiness of the hector
Word-pair:	give current equipment readiness ** *that* hector
None:	give current equipment *rating does* of the hector
Correct:	show the new definitions involving vancouver
Bigram:	show the new definitions involving vancouver
Word-pair:	show the new definitions involving vancouver
None:	show *than me* definitions involving vancouver
Correct:	show latitude and longitude of seawolf
Bigram:	show latitude and longitude of seawolf
Word-pair:	show latitude *of* longitude of seawolf
None:	show latitude ** longitude of seawolf
Correct:	show carriers that are in china sea ** and m3 on miw
Bigram:	show carriers that are in china sea *have an* m3 on miw
Word-pair:	show carriers that are in china sea and m3 on miw
None:	show carriers that *orange* ** *all* sea *ten* m3 on miw

References

[Adams 86] Adams, D. A., Bisiani, R.
The Carnegie Mellon University Distributed Speech Recognition System.
In *Speech Technology*, pages 14-23. March/April, 1986.

[Bahl 78a] Bahl, L. R., Baker, J. K., Cohen, P. S., Jelinek, F., Lewis, B. L., Mercer, R. L.
Recognition of a Continuously Read Natural Corpus.
In *IEEE International Conference on Acoustics, Speech, and Signal Processing*. April, 1978.

[Bahl 78b] Bahl, L.R., Baker, J.K., Cohen, P.S., Cole, A.G., Jelinek, F., Lewis, B.L., Mercer, R.L.
Automatic Recognition of Continuously Spoken Sentences from a Finite State Grammar.
In *IEEE International Conference on Acoustics, Speech, and Signal Processing*, pages 418-421. April, 1978.

[Bahl 80a] Bahl, L. R., Bakis, R., Cohen, P. S., Cole, A. G., Jelinek, F., Lewis, B. L., Mercer, R. L.
Further Results on the Recognition of a Continuously Read Natural Corpus.
In *IEEE International Conference on Acoustics, Speech, and Signal Processing*. April, 1980.

[Bahl 80b] Bahl, L.R., Bakis, R., Jelinek, F., Mercer, R.L.
Language-Model/Acoustic Channel Balance Mechanism.
IBM Technical Disclosure Bulletin 23(7B):3464-3465, December, 1980.

[Bahl 81a] Bahl, L. R., Bakis, R., Cohen, P. S., Cole, A. G., Jelinek, F., Lewis, B. L., Mercer, R. L.
Speech Recognition of a Natural Text Read as Isolated Words.
In *IEEE International Conference on Acoustics, Speech, and Signal Processing*. April, 1981.

[Bahl 81b] Bahl, L. R., Bakis, R., Cohen, P. S., Cole, A. G., Jelinek, F., Lewis, B. L., Mercer, R. L.
Continuous Parameter Acoustic Processing for Recognition of a Natural Speech Corpus.
In *IEEE International Conference on Acoustics, Speech, and Signal Processing*. April, 1981.

[Bahl 83a] Bahl, L. R., Jelinek, F., Mercer, R.
A Maximum Likelihood Approach to Continuous Speech Recognition.
IEEE Transactions on Pattern Analysis and Machine Intelligence PAMI-5(2):179-190, March, 1983.

[Bahl 83b] Bahl, L. R., Cole, A. G., Jelinek, F., Mercer, R. L., Nadas, A., Nahamoo, D., Picheny, M. A.
Recognition of Isolated-Word Sentences from a 5000-Word Vocabulary Office Correspondence Task.
In *IEEE International Conference on Acoustics, Speech, and Signal Processing*. April, 1983.

[Bahl 88a] Bahl. L.R., Brown, P.F., De Souza, P.V., Mercer, R.L.
Obtaining Candidate Words by Polling in a Large Vocabulary Speech Recognition System.
In *IEEE International Conference on Acoustics, Speech, and Signal Processing*. April, 1988.

[Bahl 88b] Bahl. L.R., Brown, P.F., De Souza, P.V., Mercer, R.L.
Acoustic Markov Models Used in the Tangora Speech Recognition System.
In *IEEE International Conference on Acoustics, Speech, and Signal Processing*. April, 1988.

[Baker 75a] Baker, J. K.
The DRAGON System -- An Overview.
IEEE Transactions on Acoustics, Speech, and Signal Processing ASSP-23(1):24-29, February, 1975.

[Baker 75b] Baker, J. K.
Stochastic Modeling as a Means of Automatic Speech Recognition.
PhD thesis, Computer Science Department, Carnegie Mellon University, April, 1975.

[Bakis 76] Bakis, R.
Continuous Speech Recognition via Centisecond Acoustic States.
In *91st Meeting of the Acoustical Society of America.* April, 1976.

[Baum 72] Baum, L. E.
An Inequality and Associated Maximization Technique in Statistical Estimation of Probabilistic Functions of Markov Processes.
Inequalities 3:1-8, 1972.

[Brown 83] Brown, P. F., Lee, C-H., Spohr, J. C.
Bayesian Adaptation in Speech Recognition.
In *IEEE International Conference on Acoustics, Speech, and Signal Processing*, pages 761-764. April, 1983.

[Brown 87] Brown, P.
The Acoustic-Modeling Problem in Automatic Speech Recognition.
PhD thesis, Computer Science Department, Carnegie Mellon University, May, 1987.

[Chigier 88] Chigier, B. Brennan, R.
Broad Class Network Generation Using a Combination of Rules and Statistics for Speaker Independent Continuous Speech.
In *IEEE International Conference on Acoustics, Speech, and Signal Processing.* April, 1988.

[Chow 86] Chow, Y. L., Schwartz, R., Roucos, S., Kimball, O., Price, P., Kubala, F., Dunham, M., Krasner, M., Makhoul, J.
The Role of Word-Dependent Coarticulatory Effects in a Phoneme-Based Speech Recognition System.
In *IEEE International Conference on Acoustics, Speech, and Signal Processing.* April, 1986.

[Chow 87] Chow, Y.L., Dunham, M.O., Kimball, O.A., Krasner, M.A., Kubala, G.F., Makhoul, J., Roucos, S., Schwartz, R.M.
BYBLOS: The BBN Continuous Speech Recognition System.
In *IEEE International Conference on Acoustics, Speech, and Signal Processing*, pages 89-92. April, 1987.

[Cohen 74] Cohen, P. S., Mercer, R. L.
The Phonological Component of an Automatic Speech-Recognition System.
In *Proceedings of the IEEE Symposium on Speech Recognition, Pittsburgh, PA*, pages 177-187. 1974.

[Cole 80] Cole, R. A., Rudnicky, A. I., Zue, V. W., Reddy, D. R.
Speech as Patterns on Paper.
In R. A. Cole (editor), *Perception and Production of Fluent Speech.* Lawrence Erlbaum Associates, Hillsdale, N.J., 1980.

[Cole 83] Cole, R. A., Stern, R. M., Phillips, M. S., Brill, S. M., Specker, P., Pilant, A. P.
Feature-Based Speaker Independent Recognition of English Letters.
In *IEEE International Conference on Acoustics, Speech, and Signal Processing*. October, 1983.

[Cole 86a] Cole, R. A.
Phonetic Classification in New Generation Speech Recognition Systems.
In *Speech Tech. 86*, pages 43-46. 1986.

[Cole 86b] Cole, R. A., Phillips, M., Brennan, B., Chigier, B.
The C-MU Phonetic Classification System.
In *IEEE International Conference on Acoustics, Speech, and Signal Processing*. April, 1986.

[Cravero 84] Cravero, M., Fissore, L., Pieraccini, R., Scagliola, C.
Syntax Driven Recognition of Connected Words by Markov Models.
In *IEEE International Conference on Acoustics, Speech, and Signal Processing*. April, 1984.

[Cravero 86] Cravero, M, Pieraccini, R., Raineri, F.
Definition and Evaluation of Phonetic Units for Speech Recognition by Hidden Markov Models.
In *IEEE International Conference on Acoustics, Speech, and Signal Processing*. April, 1986.

[D'Orta 87] D'Orta, P, Ferretti, M., Scarci, S.
Phoneme Classification for Real Time Speech Recognition of Italian.
In *IEEE International Conference on Acoustics, Speech, and Signal Processing*, pages 81-84. April, 1987.

[Das 83] Das, S.K.
Some Dimensionality Reduction Studies in Continuous Speech Recognition.
In *IEEE International Conference on Acoustics, Speech, and Signal Processing*, pages 292-5. April, 1983.

[Davis 80] Davis, S.B, P. Mermelstein.
Comparison of Parametric Representations of Monosyllabic Word Recognition in Continuously Spoken Sentences,
IEEE Transactions on Acoustics, Speech, and Signal Processing ASSP-28(4):357-366, August, 1980.

[Deng 88] Deng, L, Lennig, M., Gupta, V.N., Mermelstein, P.
Modeling Acoustic-Phonetic Detail in an HMM-based Large Vocabulary Speech Recognizer.
In *IEEE International Conference on Acoustics, Speech, and Signal Processing*, pages 509-512. April, 1988.

[Derouault 87] Derouault, A.-M.
Context-Dependent Phonetic Markov Models for Large Vocabulary Speech Recognition.
In *IEEE International Conference on Acoustics, Speech, and Signal Processing*, pages 360-3. April, 1987.

[Duda 73] Duda, R. O., Hart, P. E.
Pattern Classification and Scene Analysis.
John Wiley & Sons, New York, N.Y., 1973.

[Feng 88] Feng, M.W., Kubala, F., Schwartz, R.
Improved Speaker Adaptation Using Text Dependent Mappings.
In *IEEE International Conference on Acoustics, Speech, and Signal Processing.* April, 1988.

[Fisher 87] Fisher, W.M., Zue, V., Bernstein, J., Pallett, D.
An Acoustic-Phonetic Data Base.
In *113th Meeting of the Acoustical Society of America.* May, 1987.

[Furui 86] Furui, S.
Speaker-Independent Isolated Word Recognition Using Dynamic Features of Speech Spectrum.
IEEE Transactions on Acoustics, Speech, and Signal Processing ASSP-34(1):52-59, February, 1986.

[Gray 84] Gray, R.M.
Vector Quantization.
IEEE ASSP Magazine 1(2):4-29, April, 1984.

[Gupta 87] Gupta, V.N., Lennig, M., Mermelstein, P.
Integration of Acoustic Information in a Large Vocabulary Word Recognizer.
In *IEEE International Conference on Acoustics, Speech, and Signal Processing*, pages 697-700. April, 1987.

[Hamming 86] Hamming, R.W.
Coding and Information Theory.
Prentice-Hall, Englewood Cliffs NJ, 1986.

[Haton 84] Haton, J.-P.
Knowledge-based and Expert Systems in Automatic Speech Recognition.
In DeMori, R. (editor), *New Systems and Architectures for Automatic Speech Recognition and Synthesis.* Dordrecht, Reidel, Netherlands, 1984.

[Hon 88] Hon, H.W.
Personal Communication.
unpublished.
1988

[Hunt 80] Hunt, M. J., Lennig, M., Mermelstein, P.
Experiments in Syllable-Based Recognition of Continuous Speech.
In *IEEE International Conference on Acoustics, Speech, and Signal Processing*, pages 880-883. April, 1980.

[Hwang 88] Hwang, M.Y.
Personal Communication.
unpublished.
1988

[IBM 85] IBM speech recognition group.
A Real-Time, Isolated-Word, Speech Recognition System for Dictation Transcription.
In *IEEE International Conference on Acoustics, Speech, and Signal Processing.* March, 1985.

[Itakura 75] Itakura, F.
Minimum Prediction Residual Principle Applied to Speech Recognition.
IEEE Transactions on Acoustics, Speech, and Signal Processing ASSP-23(1):67-72, February, 1975.

[Jelinek 76] Jelinek, F.
Continuous Speech Recognition by Statistical Methods.
Proceedings of the IEEE 64(4):532-556, April, 1976.

[Jelinek 80] Jelinek, F., Mercer, R.L.
Interpolated Estimation of Markov Source Parameters from Sparse Data.
In E.S. Gelsema and L.N. Kanal (editor), *Pattern Recognition in Practice*, pages 381-397. North-Holland Publishing Company, Amsterdam, the Netherlands, 1980.

[Jelinek 85] Jelinek, F.
Self-Organized Language Modeling for Speech Recognition.
Unpublished.
1985

[Jelinek 87] Jelinek, F.
Personal Communication.
unpublished.
1987

[Juang 85a] Juang, B.H., Rabiner, L.R., Levinson, S.E., Sondhi, M.M.
Recent Developments in the Application of Hidden Markov Models to Speaker-Independent Isolated Word Recognition.
In *IEEE International Conference on Acoustics, Speech, and Signal Processing*, pages 9-12. April, 1985.

[Juang 85b] Juang, B. H., Rabiner, L. R.
Mixture Autoregressive Hidden Markov Models for Speech Signals.
IEEE Transactions on Acoustics, Speech, and Signal Processing ASSP-33(6):1404-13, December, 1985.

[Juang 85c] Juang, B.H., Rabiner, L.R.
A Probabilistic Distance Measure for Hidden Markov Models.
The Bell System Technical Journal 64(2):391-408, February, 1985.

[Kimball 86] Kimball, O., Price, P., Roucos, S., Schwartz, R., Kubala, F., Chow, Y.-L., Haas, A., Krasner, M., Makhoul, J.
Recognition Performance and Grammatical Constraints.
In Lee S. Baumann (editor), *Proceedings of the DARPA Speech Recognition Workshop*, pages 53-59. February, 1986.

[Klatt 72] Klatt, D.H., Stevens, K.N.
Sentence Recognition from Visual Examination of Spectrograms and Machine-Aided Lexical Searching.
In *Proceedings 1972 Conference on Speech Communication and Processing*, pages 315-318. IEEE and AFCRL, 1972.

[Klatt 86] Klatt, D.
Problem of Variability in Speech Recognition and in Models of Speech Perception.
In J.S. Perkell and D.M. Klatt (editor), *Variability and Invariance in Speech Processes*, pages 300-320. Lawrence Erlbaum Assoc, Hillsdale, N.J., 1986.

[Kubala 88] Kubala, G.F., Chow, Y., Derr, A., Feng, M., Kimball, O., Makhoul, J., Price, P., Rohlicek, J., Roucos, S., Schwartz, R., Vandegrift, J.
Continuous Speech Recognition Results of the BYBLOS System on the DARPA 1000-Word Resource Management Database.
In *IEEE International Conference on Acoustics, Speech, and Signal Processing*. April, 1988.

[Lamel 86] Lamel, L.F., Kassel, R.H., Seneff, S.
Speech Database Development: Design and Analysis of the Acoustic-Phonetic Corpus.
In Baumann, L.S. (editor), *Proceedings of the DARPA Speech Recognition Workshop*, pages 100-109. February, 1986.

[Lea 80] Lea, W.A.
Trends in Speech Recognition.
Prentice-Hall, Englewood Cliffs, NJ, 1980.

[Lee 85a] Lee, K.F.
Incremental Network Generation in Template-Based Word Recognition.
Technical Report CMU-CS-85-181, Computer Science Department, Carnegie Mellon University, December, 1985.

[Lee 85b] Lee, K.F.
Network Representation of Templates in Word Recognition.
In *The 109th Meeting of the Acoustical Society of America*. April, 1985.

[Lee 86] Lee, K.F.
Incremental Network Generation in Word Recognition.
In *IEEE International Conference on Acoustics, Speech, and Signal Processing*. April, 1986.

[Lee 87] Lee, K.F.
Towards Speaker-Independent Continuous Speech Recognition.
In *1987 NATO ASI on Speech Recognition and Dialog Understanding.* 1987.

[Lee 88a] Lee, K.F., Hon, H.W.
Large-Vocabulary Speaker-Independent Continuous Speech Recognition.
In *IEEE International Conference on Acoustics, Speech, and Signal Processing.* April, 1988.

[Lee 88b] Lee, K.F., Hon, H.W.
Speaker-Independent Phoneme Recognition Using Hidden Markov Models.
Technical Report CMU-CS-88-121, Computer Science Department, Carnegie Mellon University, Pittsburgh, PA, April, 1988.

[Lee 88c] Lee, K.F.
On Large-Vocabulary Speaker-Independent Continuous Speech Recognition.
Journal of the Eurpopean Association of Signal Processing , September, 1988.

[LeeCH 88a] Lee, C.H., Rabiner, L.R.
A Network-based Frame-synchronous Level Building Algorithm for Connected Word Recognition.
In *IEEE International Conference on Acoustics, Speech, and Signal Processing*, pages 410-413. April, 1988.

[LeeCH 88b] Lee, C.H., Soong, F.K., Juang, B.H.
A Segment Model Based Approach to Speech Recognition.
In *IEEE International Conference on Acoustics, Speech, and Signal Processing.* April, 1988.

[Lesser 75] Lesser, V. R., Fennell, R. D., Erman, L. D., Reddy, R. D.
The Hearsay II Speech Understanding System.
IEEE Transactions on Acoustics, Speech, and Signal Processing ASSP-23(1):11-24, February, 1975.

[Levinson 77] Levinson, S. E., Rosenberg, A. E., Flanagan, J. L.
Evaluation of a Word Recognition System Using Syntax Analysis.
In *IEEE International Conference on Acoustics, Speech, and Signal Processing.* April, 1977.

[Levinson 79] Levinson, S. E., Rabiner, L. R., Rosenberg, A. E., Wilpon, J. G.
Interactive Clustering Techniques for Selecting Speaker-Independent Reference Templates for Isolated Word Recognition.
IEEE Transactions on Acoustics, Speech, and Signal Processing ASSP-27(2):134-41, April, 1979.

[Levinson 83] Levinson, S. E., Rabiner, L. R., Sondhi, M. M.
An Introduction to the Application of the Theory of Probabilistic Functions on a Markov Process to Automatic Speech Recognition.
The Bell System Technical Journal 62(4), April, 1983.

[Linde 80] Linde, Y., Buzo, A., Gray, R.M.
An Algorithm for Vector Quantizer Design.
IEEE Transactions on Communication COM-28(1):84-95, January, 1980.

[Lippmann 87] Lippmann, R.P., Martin, E.A., Paul, D.P.
Multi-Style Training for Robust Isolated-Word Speech Recognition.
In *IEEE International Conference on Acoustics, Speech, and Signal Processing*, pages 705-8. April, 1987.

[Lowerre 76] Lowerre, B. T.
The HARPY Speech Recognition System.
PhD thesis, Computer Science Department, Carnegie Mellon University, April, 1976.

[Lowerre 77] Lowerre, B. T.
Dynamic Speaker Adaptation in the Harpy Speech Recognition System.
In *IEEE International Conference on Acoustics, Speech, and Signal Processing*. April, 1977.

[Lowerre 80] Lowerre, B.T., Reddy, D.R.
The Harpy Speech Understanding System.
Trends in Speech Recognition.
Prentice-Hall, Englewood Cliffs, NJ, 1980.

[Lucassen 83] Lucassen, J.M.
Discovering Phonemic Baseforms: an Information Theoretic Approach.
Research Report RC 9833, IBM, February, 1983.

[Makhoul 85] Makhoul, J., Roucos, S., Gish, H.
Vector Quantization in Speech Coding.
Proceedings of the IEEE 73(11):1551-1588, November, 1985.

[Markel 76] Markel, J. D., Gray, A. H.
Linear Prediction of Speech.
Springer-Verlag, Berlin, 1976.

[Meilijson 87] Meilijson, I.
A Fast Improvement to the EM Algorithm on its Own Terms.
Forthcoming in the Journal of the Royal Statistical Society.
1987

[Merialdo 87] Merialdo, B.
Speech Recognition With Very Large Size Dictionary.
In *IEEE International Conference on Acoustics, Speech, and Signal Processing*, pages 364-7. April, 1987.

[Murveit 88] Murveit, H., Weintraub, M.
Speaker-Independent Connected-Speech Recognition Using Hidden Markov Models.
In *IEEE International Conference on Acoustics, Speech, and Signal Processing*. April, 1988.

[Myers 81] Myers, C.S., Rabiner, L.R.
Connected Digit Recognition Using a Level Building DTW Algorithm.
ASSP ASSP-29(3):351-363, June, 1981.

[Nadas 81] Nadas, A., Mercer, R. L., Bahl, L. R., Bakis, R., Cohen, P. S., Cole, A. G., Jelinek, F., Lewis, B. L.
Continuous Speech Recognition with Automatically Selected Acoustic Prototypes Obtained by Either Bootstrapping or Clustering.
In *IEEE International Conference on Acoustics, Speech, and Signal Processing*. April, 1981.

[Nag 86] Nag, R., Austin, S.C., Fallside, F.
Using Hidden Markov Models to Define Linguistic Units.
In *IEEE International Conference on Acoustics, Speech, and Signal Processing*, pages 2239-42. April, 1986.

[Ney 87] Ney, H., Mergel, D., Noll, A., Paeseler, A.
A Data-Driven Organization of the Dynamic Programming Beam Search for Continuous Speech Recognition.
In *IEEE International Conference on Acoustics, Speech, and Signal Processing*, pages 833-836. April, 1987.

[Ney 88] Ney, H., Noll, A.
Phoneme Modelling Using Continuous Mixture Densities.
In *IEEE International Conference on Acoustics, Speech, and Signal Processing*, pages 437-440. April, 1988.

[Nilsson 80] Nilsson, N.J.
Principles of Artificial Intelligence.
Tioga Publishing Co., Palo Alto, CA, 1980.

[Nishimura 87] Nishimura, M., Toshioka, K.
HMM-Based Speech Recognition Using Multi-Dimensional Mutli-Labeling.
In *IEEE International Conference on Acoustics, Speech, and Signal Processing*, pages 1163-6. April, 1987.

[Noll 87] Noll, A. Ney, H.
Training of Phoneme Models in a Sentence Recognition System.
In *IEEE International Conference on Acoustics, Speech, and Signal Processing*, pages 1277-80. April, 1987.

[Oppenheim 72] Oppenheim, A. V., Johnson, D. H.
Discrete Representation of Signals.
The Proceedings of the IEEE 60(6):681-691, June, 1972.

[Paul 86] Paul, D. B., Lippmann, R. P., Chen, Y., Weinstein, C.
Robust HMM-Based Techniques for Recognition of Speech Produced under Stress and in Noise.
In *Speech Tech.* April, 1986.

[Paul 88] Paul, D.B., Martin, E.A.
Speaker Stress-Resistant Continuous Speech Recognition.
In *IEEE International Conference on Acoustics, Speech, and Signal Processing*. April, 1988.

[Picheny 88] Picheny, M.
Personal Communication.
unpublished.
1988

[Polifroni 88] Polifroni, J.
Personal Communication.
unpublished.
1988

[Price 88] Price, P.J., Fisher, W., Bernstein, J., Pallett, D.
A Database for Continuous Speech Recognition in a 1000-Word Domain.
In *IEEE International Conference on Acoustics, Speech, and Signal Processing.* April, 1988.

[Rabiner 79] Rabiner, L. R., Levinson, S. E., Rosenberg, A. E., Wilpon, J. G.
Speaker-Independent Recognition of Isolated Words Using Clustering Techniques.
IEEE Transactions on Acoustics, Speech, and Signal Processing ASSP-27(4):336-349, August, 1979.

[Rabiner 81] Rabiner, L. R., Wilpon, J. G.
Isolated Word Recognition Using a Two-Pass Pattern Recognition Approach.
In *IEEE International Conference on Acoustics, Speech, and Signal Processing*, pages 724-7. March, 1981.

[Rabiner 83] Rabiner, L. R., Levinson S. E., Sondhi, M. M.
On the Application of Vector Quantization and Hidden Markov Models to Speaker-Independent, Isolated Word Recognition.
The Bell System Technical Journal 62(4):1075-1105, April, 1983.

[Rabiner 84] Rabiner, L. R., Pan, K. C., Soong, F. K.
On the Performance of Isolated Word Speech Recognizers Using Vector Quantization and Temporal Energy Contours.
AT&T Bell Laboratories Technical Journal 63(7):1245-1260, September, 1984.

[Rabiner 85] Rabiner, L. R., Juang, B. H., Levinson, S. E., Sondhi, M. M.
Recognition of Isolated Digits Using Hidden Markov Models With Continuous Mixture Densities.
AT&T Technical Journal 64(6):1211-33, July-August, 1985.

[Rabiner 86] Rabiner, L.R., Juang, B.H.
An Introduction to Hidden Markov Models.
IEEE ASSP Magazine 3(1):4-16, January, 1986.

[Rabiner 88a] Rabiner, L.R., Wilpon, J.G., Soong, F.K.
High Performance Connected Digit Recognition Using Hidden Markov Models.
In *IEEE International Conference on Acoustics, Speech, and Signal Processing*. April, 1988.

[Rabiner 88b] Rabiner, L.R.
A Tutorial on Hidden Markov Models and Selected Applications in Speech Recognition.
IEEE Proceedings , 1988.

[Reddy 77] Reddy, D. R.
Speech Understanding Systems: Summary of Results of the Five-Year Research Effort at Carnegie Mellon University.
Internal Document.
August, 1977

[Reddy 83] Reddy, D.R., Zue, V.
Recognizing Continuous Speech Remains an Illusive Goal.
IEEE Spectrum :84-87, November, 1983.

[Richter 86] Richter, A.G.
Modeling of Continuous Speech Observations.
In *Advances in Speech Processing Conference, IBM Europe Institute*. July, 1986.

[Rosenberg 83] Rosenberg, A. E., Rabiner, L. R., Wilpon, J., Kahn, D.
Demisyllable-Based Isolated Word Recognition System.
IEEE Transactions on Acoustics, Speech, and Signal Processing ASSP-31(3):713-726, June, 1983.

[Roucos 87] Roucos, S., Dunham, M.O.
A Stochastic Segment Model for Phoneme-Based Continuous Speech Recognition.
In *IEEE International Conference on Acoustics, Speech, and Signal Processing*, pages 73-76. April, 1987.

[Roucos 88] Roucos, S., Ostendorf, M., Gish, H., Derr, A.
Stochastic Segment Modeling Using the Estimate-Maximize Algorithm.
In *IEEE International Conference on Acoustics, Speech, and Signal Processing*. April, 1988.

[Rudnicky 87] Rudnicky, A., Baumeister, L., DeGraaf, K., Lehmann, E. The Lexical Access Component of the CMU Continuous Speech Recognition System. In *IEEE International Conference on Acoustics, Speech, and Signal Processing*. April, 1987.

[Ruske 82] Ruske, G. Auditory Perception and Its Application to Computer Analysis of Speech. In C. Y. Suen and R. De Mori (editor), *Auditory Signals*. Volume II: *Computer Analysis and Perception*. CRC Press, Boca Raton, FL, 1982.

[Russell 85] Russell, M.J., Moore, R.K. Explicit Modeling of State Occupancy in Hidden Markov Models for Automatic Speech Recognition. In *IEEE International Conference on Acoustics, Speech, and Signal Processing*, pages 5-8. April, 1985.

[Schwartz 80] Schwartz, R., Klovstad, J., Makhoul, J., Sorensen, J. A Preliminary Design of a Phonetic Vocoder Based on a Diphone Model. In *IEEE International Conference on Acoustics, Speech, and Signal Processing*, pages 32-35. April, 1980.

[Schwartz 84] Schwartz, R. M., Chow, Y. L., Roucos, S., Krasner, M., Makhoul, J. Improved Hidden Markov Modeling of Phonemes for Continuous Speech Recognition. In *IEEE International Conference on Acoustics, Speech, and Signal Processing*. April, 1984.

[Schwartz 85] Schwartz, R., Chow, Y., Kimball, O., Roucos, S., Krasner, M., Makhoul, J. Context-Dependent Modeling for Acoustic-Phonetic Recognition of Continuous Speech. In *IEEE International Conference on Acoustics, Speech, and Signal Processing*. April, 1985.

[Schwartz 87] Schwartz, R., Chow, Y., Kubala, F. Rapid Speaker Adaptation Using a Probabilistic Spectral Mapping. In *IEEE International Conference on Acoustics, Speech, and Signal Processing*. April, 1987.

[Shikano 85] Shikano, K.
Evaluation of LPC Spectral Matching Measures for Phonetic Unit Recognition.
Technical Report, Computer Science Department, Carnegie Mellon University, May, 1985.

[Shikano 86a] Shikano, K., Lee. K, Reddy, D. R.
Speaker Adaptation through Vector Quantization.
In *IEEE International Conference on Acoustics, Speech, and Signal Processing.* April, 1986.

[Shikano 86b] Shikano, K., Lee, K. Reddy, D. R.
Speaker Adaptation through Vector Quantization.
Technical Report CMU-CS-86-102, Computer Science Department, Carnegie Mellon University, December, 1986.

[Shikano 86c] Shikano, K.
Evaluation of LPC Spectral Matching Measures for Phonetic Unit Recognition.
Technical Report, Computer Science Department, Carnegie Mellon University, 1986.

[Shikano 86d] Shikano, K.
Text-Independent Speaker Recognition Experiments using Codebooks in Vector Quantization.
Technical Report, Computer Science Department, Carnegie Mellon University, 1986.

[Shipman 82] Shipman, D.W., Zue, V.W.
Properties of Large Lexicons: Implications for Advanced Isolated Word Recognition Systems.
In *IEEE International Conference on Acoustics, Speech, and Signal Processing*, pages 546-549. April, 1982.

[Stern 83] Stern, R. M., Lasry, M. J.
Dynamic Speaker Adaptation for Isolated Letter Recognition Using MAP Estimation.
In *IEEE International Conference on Acoustics, Speech, and Signal Processing.* April, 1983.

[Sugawara 85] Sugawara, K., Nishimura, M., Toshioka, K., Okochi, M., Kaneko, T.
Isolated Word Recognition Using Hidden Markov Models.
In *IEEE International Conference on Acoustics, Speech, and Signal Processing.* April, 1985.

[Thompson 87] Thompson, H.S., Laver, J.D.
The Alvey Speech Demonstrator - Architecture, Methodology, and Progress to Date.
In *Proceedings of Speech Tech.* 1987.

[TI 87] TI Speech Recognition Group.
TI Speech Recognition Technology Development.
In *DARPA Speech Recognition Workshop.* October, 1987.

[Tohkura 86] Tohkura, Y.
A Weighted Cepstral Distance Measure for Speech Recognition.
In *IEEE International Conference on Acoustics, Speech, and Signal Processing.* April, 1986.

[Viterbi 67] Viterbi, A. J.
Error Bounds for Convolutional Codes and an Asymptotically Optimum Decoding Algorithm.
IEEE Transactions on Information Theory IT-13(2):260-269, April, 1967.

[Waibel 86] Waibel, A. H.
Prosody and Speech Recognition.
PhD thesis, Computer Science Department, Carnegie Mellon University, October, 1986.

[Wilpon 82] Wilpon, J. G., Rabiner, L. R., Bergh, A.
Speaker-Independent Isolated Word Recognition Using a 129-Word Airline Vocabulary.
The Journal of the Acoustical Society of America 72(2):390-396, August, 1982.

[Zue 85] Zue, V. W.
The Use of Speech Knowledge in Automatic Speech Recognition.
Proceedings of the IEEE 73(11):1602-1615, November, 1985.

[Zwicker 61] Zwicker, E.
Subdivision of the Audible Frequency Range into Critical Bands (Frequenzgruppen).
Journal of the Acoustical Society of America 33:248, February, 1961.

Index